MULTIPLE CHOICE QUESTIONS TO PREPARE FOR THE AP CALCULUS AB EXAM

(second edition)

By Rita Korsunsky

www.mathboat.com

ISBN: 1481283235
ISBN-13: 9781481283236

PREFACE

I am delighted to offer this book, titled "MULTIPLE CHOICE QUESTIONS TO PREPARE FOR THE AP CALCULUS AB EXAM ".

This book consists of two parts. The first part contains five sample multiple-choice exams. Section I Part A consists of 30 multiple-choice questions to be answered without the use of a calculator. Section 1 part B consists of 15 multiple-choice questions where a graphing calculator could be used. The second part of this book contains a comprehensive list of definitions, formulas and theorems, and tips for the AP test.

The sole purpose of this book is to help you prepare for the AP Calculus AB exam. On the exam, you will have to respond to both multiple-choice and free-response questions. It is my belief that extensive practice with multiple-choice questions will help to prepare you well for both sections.

Do not forget to practice for the free response part of the test. I strongly suggest that you practice by answering the free response questions from past AP tests. These are available for free on the College Board's website.

Completing the multiple choice tests offered in this book will enable you to learn your strengths and weaknesses. You will find out what material you need to review. If you focus on learning, re-learning, and reviewing this material, you will greatly improve your performance on the test.

There is no substitute for practice. Taking multiple practice tests will teach you what your problem areas are, and will increase your confidence on the day of the exam.

I would like to thank all of my past, present, and future students for inspiring me to publish this Book. A special thanks to Lillian Li for hand drawing the exquisite book cover and Travis Chen for helping with publishing procedure. I also wish to thank my sons David and Boris for their technical assistance and support. A special debt of gratitude is due to my husband Alex for his continual support and help every step of the way.

Please send all questions and concerns to:

captain@mathboat.com

Sincerely,

Rita Korsunsky

If you wish to order the solutions with step-by-step explanations to each and every problem made in a form of PowerPoint presentation, please visit www.mathboat.com

TABLE OF CONTENTS

Examination I

Section I Part A

Directions: Solve each of the following problems, using the space provided. Choose the best answer. Do not spend too much time on any one problem. Calculators may NOT be used on this part of the exam.

In this test: Unless otherwise specified, the domain of a function is assumed to be the set of all real numbers x for which $f(x)$ is a real number.

1. $\lim_{h \to 0} \dfrac{\sin\left(\dfrac{\pi}{2}+h\right)-\sin\left(\dfrac{\pi}{2}\right)}{h}$ is

(A) $\dfrac{\sqrt{2}}{2}$

(B) 1

(C) 0

(D) nonexistent

Answer____________________

2. The position of an object moving along the line at time t is described by the function $s(t)=-(t^2-2t+4)(t^3-2t)$. What is the velocity of the object at time $t=1$?

(A) -3

(B) -1

(C) 1

(D) 3

Answer ____________________

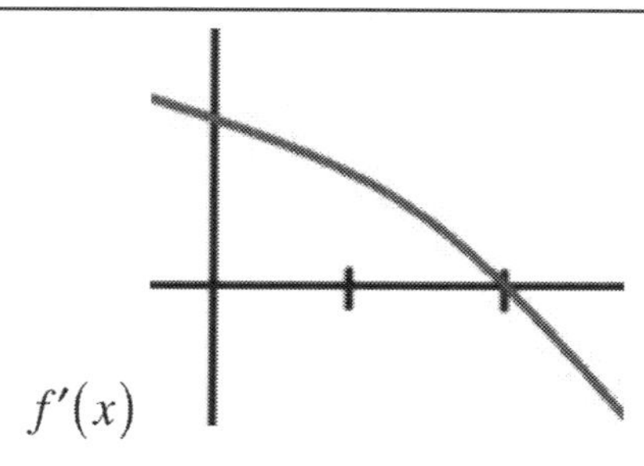

3. The graph of $f'(x)$ is shown above. The function $f(x)$ is twice differentiable.
If $f(2)=4$, which of the following is true?

$(A)\, f''(2) < f(2) < f'(2)$

$(B)\, f''(2) < f'(2) < f(2)$

$(C)\, f'(2) < f''(2) < f(2)$

$(D)\, f(2) < f''(2) < f'(2)$

Answer________________

4. If the function f defined by $f(x)=\begin{cases} 1+e^{-2x}, & 0 \le x \le b \\ 1+e^{2x-12}, & b < x \le 6 \end{cases}$

is continuous for all values of x on the interval $[0,6]$, which of the following is the value of b?

$(A)\,1$ $(B)\,2$ $(C)\,3$ $(D)\,4$

Answer________________

5. Given $f(x) = x^3 - 6x^2 + 12x - 8$ and $g(x)$ is its inverse, what is $g'(-8)$?

(A) $\frac{1}{8}$

(B) 12

(C) $\frac{1}{12}$

(D) 8

Answer____________________

6. A rubber band in the shape of a circle is being evenly stretched, and its radius is increasing at a constant rate of 1.5 cm/sec. At what rate (in cm/sec) is the circumference of the rubber band changing when its radius is 6 centimeters?

(A) 2π

(B) 3π

(C) 6π

(D) 18π

Answer ____________________

7. $\lim_{x\to 0}\frac{\ln(\cos x)-e^{x}+x+1}{\tan x}=$

$(A)\ 0$ $(B)\ 1$ $(C)\ \pi$ $(D)\ 2$

Answer________________

8. The volume of the solid formed by revolving the region bounded by the graph of $2y=x^2+4$ and $y=x^2$ about $y=4$ is given by which of the following integrals?

$(A)\pi\int_{-2}^{2}\left(2-\frac{x^2}{2}\right)^2 dx$

$(B)\pi\int_{-2}^{2}\left(\left(x^2\right)^2-\left(\frac{x^2}{2}+2\right)^2\right)dx$

$(C)\pi\int_{-2}^{2}\left(\left(4-x^2\right)^2-\left(2-\frac{x^2}{2}\right)^2\right)dx$

$(D)\pi\int_{-2}^{2}\left(\left(4-x^2\right)^2-\left(\frac{x^2}{2}\right)^2\right)dx$

Answer ________________

9. The rate at which water is being pumped into a water tank is modeled by the function $f(t) = -\frac{1}{4}t^3 + \frac{3}{2}t^2 + 3$. For the time interval $1 \le t \le 7$, at what time t is the water being pumped at the fastest rate?

$(A)\ t = 2 \quad (B)\ t = 3 \quad (C)\ t = 4 \quad (D)\ t = 6$

Answer____________________

10. If $e^{xy+1} = 3$, what is $\frac{dy}{dx}$ at $x = 1$?

$(A)\ \frac{1}{\ln 3} \quad (B)\ 1 - \ln 3 \quad (C)\ \ln 3 - 1 \quad (D)\ \ln 3$

Answer ____________________

11. A line perpendicular to the tangent line to the curve at the point of tangency is called a normal line to the curve at that point. Which of the following is the equation of the normal line to the curve $f(x)=\sqrt{x^3+3}$ at the point on the curve where $x=1$?

$(A)\ 3y+4x=10$ $(B)\ 4y-3x=5$ $(C)\ 3y-4x=2$ $(D)\ 3y+4x=14$

Answer____________________

12. $\int_1^2 \left(1+\frac{1}{x}\right)^{-2}\left(\frac{1}{x^2}\right)dx=$

$(A)\ \frac{1}{2}$ $(B)\ \frac{1}{6}$ $(C)\ \frac{1}{12}$ $(D)\ \frac{1}{3}$

Answer ____________________

13. If $\lim_{x\to 3} \frac{f(x)}{2x-6} = f'(3) = 0$, which of the following must be true?

I. $(3,0)$ is a critical point of $f(x)$

II. $f(x)$ has a local maximum at $x = 3$

III. $f(x)$ is continuous at $x = 3$

(A) I only

(B) III only

(C) I and III only

(D) I, II, and III

Answer____________________

14. The function f shown on the right is defined on the closed interval $[-4,5]$. If the function p is defined by $p(x) = f(-2x^3)$, what is the slope of the line tangent to the graph of p at the point where $x = -1$?

(A) 12 (B) -12 (C) 6 (D) -2

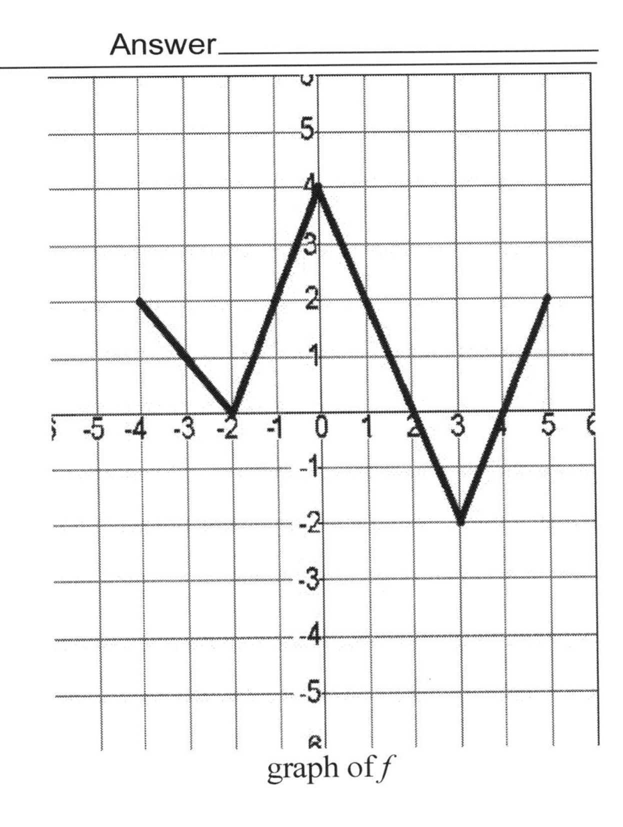

graph of f

Answer ____________________

15. What is the absolute maximum of the function $f(x)=\frac{1}{2}x^4-\frac{4}{3}x^3+2$ on the interval $[-1, 3]$?

$(A)3$ $(B)\frac{23}{6}$ $(C)\frac{13}{2}$ $(D)\frac{17}{2}$

Answer____________________

16. If $f'(x)=12x^2-6x+3$ and $f(1)=15$, what is $f(x)$?

$(A)4x^3-3x^2+3x+1$ $(B)4x^3-3x^2+3x+11$ $(C)4x^3-6x^2+3x+1$
$(D)12x^3-6x-12$

Answer ____________________

17. Let the function $f(x)$ be continuos for all x. The figure below shows the graph of $f''(x)$.

Which of the following statements about $f(x)$ or $f'(x)$ must be true?

I. f has a relative maximum in the open interval $c < x < d$.

II. f is concave up for $x < a$ and $b < x < c$.

III. f' has a relative minimum in the open interval $a < x < c$

(A) II only (B) III only (C) II and III only (D) I, II, and III

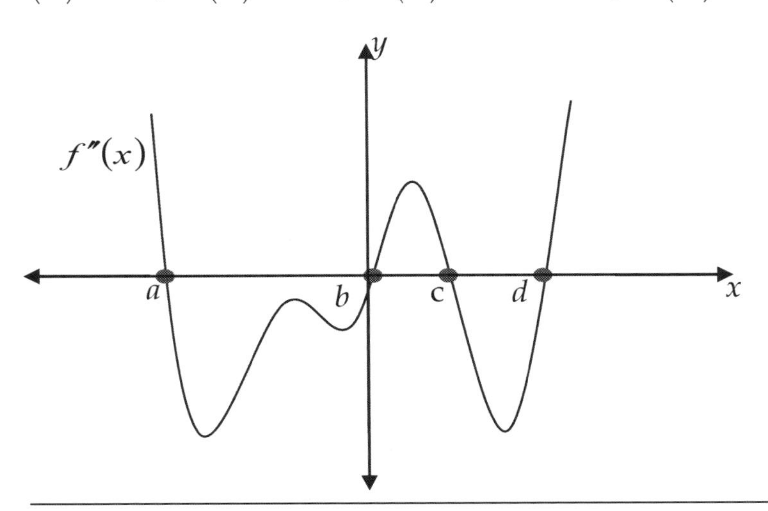

Answer____________________

18. What is the 57th derivative of $y = \cos 7x$?

$(A)\ -7^{57}\sin 7x$ $(B)\ 7^{57}\sin 7x$ $(C)\ -7^{57}\cos 7x$ $(D)\ 7^{57}\cos 7x$

Answer ____________________

19. What is the slope of the tangent line to the curve $\left(x^3+y\right)^2-2a^2xy=b^2$ at the point $P(0,1)$ when $a=\sqrt{2}$ and $b=1$?

$(A)-2$

$(B)-\frac{1}{2}$

$(C)\ 2$

$(D)\frac{1}{2}$

Answer______________________

20. A circle is inscribed in a square. The area of the circle is increasing at a constant rate of 15π in^2/sec. As the circle expands, the square expands to keep the circle inscribed. At what rate is the area of the square increasing in in^2/sec?

$(A)\ 30\quad (B)\ \frac{60}{\pi}\quad (C)\ \frac{120}{\pi}\quad (D)\ 60$

Answer ______________________

21. A boy runs on a straight road for $0 \le t \le 8$ seconds. The graph below, which consists of two line segments, shows the velocity, in meters per second, of the boy. What is the total distance, in meters, run by the boy over the time interval $0 \le t \le 8$ seconds?

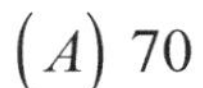

(A) 70

(B) 70.5

(C) 71

(D) 71.5

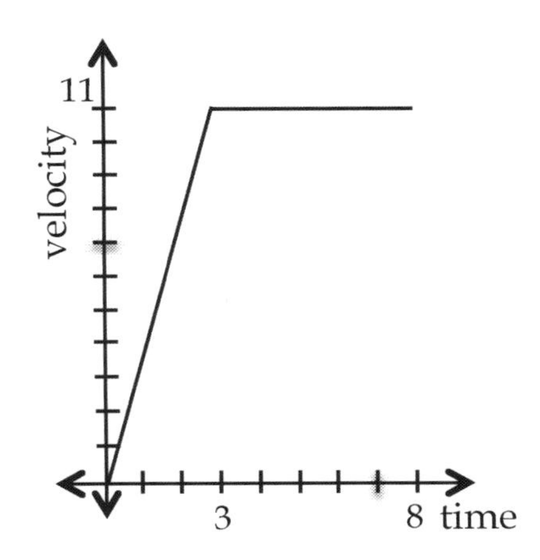

Answer________________________

22. If n is a positive integer, then $\lim_{n\to+\infty} \frac{1}{n}\left[\frac{1}{1+\left(\frac{n+1}{n}\right)^2}+\frac{1}{1+\left(\frac{n+2}{n}\right)^2}+\ldots+\frac{1}{1+\left(\frac{n+n}{n}\right)^2}\right]$ could be expressed as

$(A)\ \int_1^2 \frac{1}{x\left(x^2+1\right)}\,dx$

$(B)\ \int_1^2 \frac{1}{x^2+1}\,dx$

$(C)\ \int_0^2 \frac{1}{x\left(x^2+1\right)}\,dx$

$(D)\ \int_0^2 \frac{1}{x^2+1}\,dx$

Answer ________________________

23. Given the function $f(x)=x^2+4x-1$, for which of the following values of c on the open interval $(0,5)$ will the conclusion of the Mean Value Theorem be satisfied for the function $f(x)$?

$(A)\ \frac{9}{2}$

$(B)\ 4$

$(C)\ 1$

$(D)\ \frac{5}{2}$

Answer____________________

24. Shown on the right is the slope field for which differential equation?

$(A)\ \frac{dy}{dx}=2x$

$(B)\ \frac{dy}{dx}=2x-4$

$(C)\ \frac{dy}{dx}=4-2x$

$(D)\ \frac{dy}{dx}=x+y$

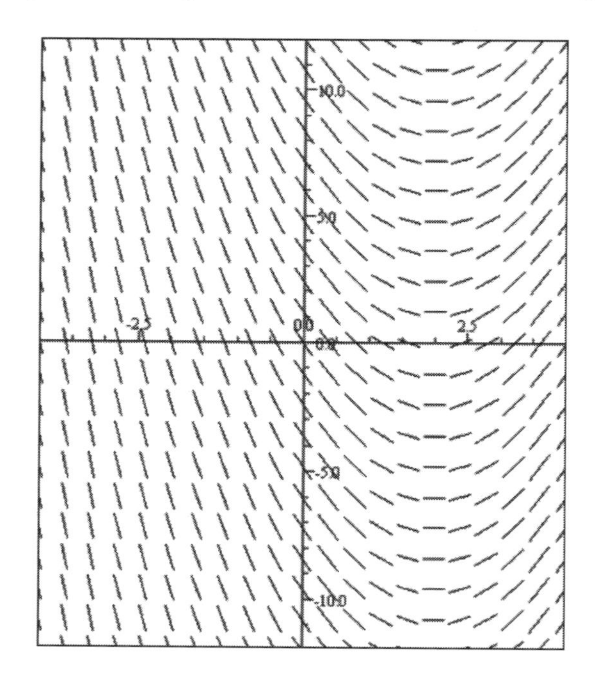

Answer____________________

25. $\lim_{x\to 2}\dfrac{\sqrt{x^2+5}-3}{x^2-4}$ is

(A) nonexistent (B) 0 (C) $\frac{1}{6}$ (D) $\frac{1}{3}$

Answer____________________

26. What are the interval(s) where the function $f(x)=\frac{8}{3}x^3-6x^2-36x+7$ is increasing?

(A) $\left[-\frac{3}{2},3\right]$

(B) $(-\infty,1]$ and $[3,\infty)$

(C) $\left(-\infty,-\frac{3}{2}\right]$ and $[3,\infty)$

(D) $\left(-\infty,-\frac{3}{2}\right]$ and $[1,\infty)$

Answer ____________________

27. Which of the following is the equation of the function $f(x)$ if its second derivative, $f''(x)=2\sin x-3\cos x$ and $f(0)=7, f'(0)=3$?

(A) $f(x)=2\cos x+3\sin x+7x+5$ (B) $f(x)=-2\sin x+3\cos x+5x+4$

(C) $f(x)=-2\sin x+3\cos x+5x+7$ (D) $f(x)=3\sin x-2\cos x-3x+9$

Answer____________________

28. If $f'(x)=\lim\limits_{h\to 0}\dfrac{(x+h)^3-x^3}{h}$, which of the following must be true:

I. $f'(x)\le 0$ for all x

II. $f'(x)\ge 0$ for all x

III. $f(x)$ is concave down when $x\ge 0$

(A) I only

(B) II only

(C) III only

(D) II and III only

Answer ____________________

29. Using the table below, what is $\left(\frac{f}{g}\right)'(3)$?

(A) $-\frac{2}{3}$

(B) $-\frac{2}{27}$

(C) $\frac{2}{3}$

(D) $\frac{2}{27}$

$f(3)$	$g(3)$	$f'(3)$	$g'(3)$
7	-9	-4	6

Answer____________________________

30. The graph of f'', the second derivative of f, is the line shown in the figure below. If $f'(0)=2.5$, then $f'(2)=$

(A) 9.5

(B) 10.5

(C) 11.0

(D) 11.5

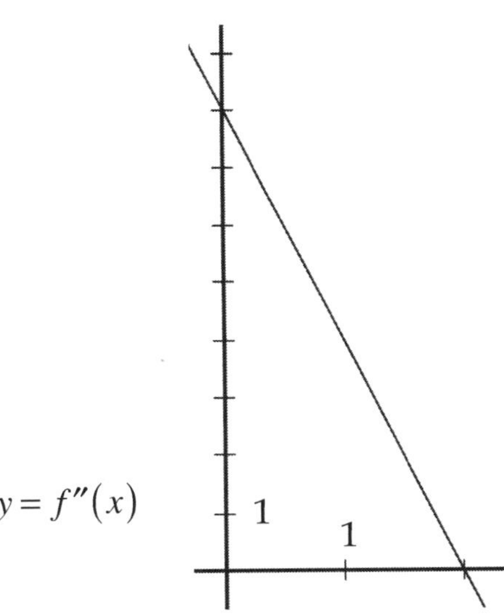

Answer ____________________________

Section I Part B

Directions: Solve each of the following problems, using the space provided. Choose the best answer. Do not spend too much time on any one problem. A graphing calculator is required for some questions on this part of the exam.

In this test:

1. The exact numerical value of the correct answer does not always appear among the choices given. Then select from among the choices the number that best approximates the exact numerical value.
2. Unless otherwise specified, the domain of a function is assumed to be the set of all real numbers x for which $f(x)$ is a real number.

31. Let R be the region in the first quadrant bounded by the graph of $y=-4x^2+2kx$ and the x-axis. If the area of the region R is 18, then the value of k is

$(A)3$ $(B)6$ $(C)9$ $(D)18$

Answer________________

32. The composition of two functions, $h(x)=f(g(x))$ is a differentiable function. Using the tables below what is the approximate value of the derivative of $h(x)$ at $x=1.35$?

(A) 15.88 (B) 10.72 (C) 3.97 (D) 13.29

x	2.5	2.7	2.9
$f'(x)$	3.85	3.97	4.03

x	1.30	1.35	1.40
$g(x)$	2.5	2.7	2.9

Answer________________

33. If $f(2) = 8$ and $f'(x) = \dfrac{\cos(1-x^2)}{x^2+\sqrt{x}}$, then $f(7) \approx ?$

(A) 8.170

(B) 0.170

(C) 7.982

$(D) -8.018$

Answer____________________

34. The function $f(x) = \left|\sin^2 x - \cos x \sin\left(\dfrac{\pi x}{2}\right)\right|$ is given on the interval $(0,5)$. Approximately at which x-coordinates is f not differentiable?

$(A)\{0.9, 2.4, 3.7\}$ (B) $\{0.4, 1.7, 3.1, 4.5\}$ (C) $\{2.4, 4.5\}$ (D) $\{0.9, 1.7, 4.5\}$

Answer ____________________

35. The function $f(x)=\sin x$ is defined on the interval $[0,b]$. The line tangent to the graph of $f(x)$ at $x=1.531$ is parallel to the segment connecting the endpoints of $[0,b]$. What is the value of b?

(A) 2.704

(B) 4.787

(C) 3.567

(D) 3.021

Answer____________________

36. Water is being pumped continuously from a water pool at a rate proportional to the amount of water left in the pool; that is, $\frac{dy}{dt}=ky$, where y is the amount of water left in the pool at any time t. Initially there were 500,000 gallons of water in the pool, and 10 days later there were 100,000 gallons left. What is the equation for y, the amount of water remaining in the pool at any time t?

$(A)\ y(t)=500{,}000e^{\frac{1}{5}t}$

$(B)\ y(t)=500{,}000\left(\frac{1}{10}\right)^{\frac{t}{10}}$

$(C)\ y(t)=500{,}000\cdot\left(\frac{1}{5}\right)^{\frac{t}{10}}$

$(D)\ y(t)=500{,}000e^{10t}$

Answer ____________________

37. Two bikers start at the same place and same time. The first biker rides east at a constant velocity of 5 miles per hour and the second biker rides north at a constant velocity of 12 miles per hour. Approximately how fast is the distance between them changing after 3 hours?

(A) 7mph

(B) 12mph

(C) 13mph

(D) 17mph

Answer____________________

38. Let f be the function whose derivative is $f'(x)=\frac{1+e^x}{x^2}$ and whose graph passes through the point (3,6). What is the approximate value of $f(3.1)$ if $\int_3^{3.1} f'(x)\,dx = 0.2377$?

(A) 6.238 (B) 2.414 (C) 6.1 $(D)-5.762$

Answer ____________________

39. What is the average value of the function $G(x)=30-\frac{x^8}{2,500,000}$ on the interval where $G(x)\geq 0$?

(A) 28.473

(B) 27.134

(C) 26.667

(D) 25.431

Answer______________________

40. The region bounded by the graphs of the equations $y=5$ and $5x^4-y=3x^2-3$ is revolved around the $x-$axis. What is the approximate volume of the resulting solid?

(A) 31.416

(B) 68.793

(C) 81.274

(D) 98.696

Answer ______________________

41. What is the best approximation for the area of the region bounded by $y = \cos^2 x$ and $y = \sin^2 x$ between $x = \frac{\pi}{4}$ and $x = \frac{3\pi}{4}$?

(A) 1

(B) 1.57

(C) 1.66

(D) 2

Answer______________________

42. At which x-coordinate does the function $f(x) = 2\sin x + 4x$ have a point of inflection on the interval $(0, 2\pi)$?

(A) $\frac{\pi}{2}$

(B) $\frac{3\pi}{4}$

(C) $\frac{3\pi}{2}$

(D) π

Answer ______________________

43. A company is producing parts to sell to an auto parts store. The store buys in batches of 100, and pays the company $800 for each batch of 100 sold. The cost for the company to make x parts is modeled by the function $c(x)=\frac{x^{\frac{3}{2}}}{12}+5x-10$. To the nearest whole dollar,what is the maximum possible profit for the company?

(A) 578

(B) 585

(C) 586

(D) 600

Answer____________________

44. The base of a solid is the region in the fourth quadrant enclosed by the graph of $y=x^2-9$ and the coordinate axes. If every cross section perpendicular to the x-axis is a square, then the volume of the solid is

(A)64.8 (B)129.6 (C)171.0 (D)194.4

Answer ____________________

45. If $\int_1^7 f(x)\,dx = 6$ and $\int_7^1 g(x)\,dx = 12$ and $\int_1^7 h(x)\,dx = \int_1^5 \left(2f(x) - g(x)\right)dx + \int_7^5 \left(g(x) - 2f(x)\right)dx$,
what is the average value of $h(x)$ on the interval $1 \le x \le 7$?

(A) 1 (B) 2 (C) 4 (D) 12

Answer____________________

Examination II

Section I Part A

Directions: Solve each of the following problems, using the space provided. Choose the best answer. Do not spend too much time on any one problem. Calculators may NOT be used on this part of the exam.

In this test: Unless otherwise specified, the domain of a function is assumed to be the set of all real numbers x for which $f(x)$ is a real number.

1. Let f be defined as follows, where $a \neq 0$.

$$f(x)=\begin{cases} \dfrac{x^3-a^3}{x^2-a^2}, & \text{for } x \neq a \\ 1, & \text{for } x=a \end{cases}$$

Which of the following must be true about f ?

I. $\lim_{x \to a} f(x)$ exists

II. $f(a)$ exists

III. $f(x)$ is continuous at $x=a$

(A) I only (B) II only (C) I and II only (D) I, II, and III

Answer ____________

2. The slope of a curve at point (x,y) is defined as $\lim_{h \to 0} \dfrac{(x+h)^3+(x+h)^2-x^3-x^2}{h}$. Which of the following is the equation of the tangent to this curve at $x=1$?

$(A)\ y=5x-2$

$(B)\ y=3x-9$

$(C)\ y=5x-3$

$(D)\ y=3x-6$

Answer ____________

3. $\int x \cdot (2x^2+1)^4 dx =$

(A) $\frac{1}{4}(2x^2+1)^5 + C$

(B) $\frac{1}{4}(2x^2+1)^4 + C$

(C) $\frac{1}{20}(2x^2+1)^5 + C$

(D) $\frac{1}{20}(2x^2+1)^6 + C$

Answer ______________________

4. If $f(x) = g(h(x))$ and if $h(2) = 5$, $h'(2) = -5$, $g'(5) = 3$ which of the following is $f'(2)$?

(A) 3

(B) -15

(C) 15

(D) -3

Answer ______________________

5. Which of the following is y', the first derivative of the function $y = f(x)$ if $x^2 y + \sec y = 8$?

$(A)\ -2xy \cdot (x^2 \sec y \tan y)$ $(B)\ \dfrac{x^2 y}{\sec y \tan y}$ $(C)\ \dfrac{-2xy}{x^2 - \sec y \tan x}$ $(D)\ \dfrac{-2xy}{x^2 + \sec y \tan y}$

Answer ______________

6. If $f(x) = 6x^2 + 4x$, what is the number c in the interval $[0,4]$ such that the line tangent to the graph of $f(x)$ at the point $x = c$ is parallel to the line drawn through the endpoints of the interval?

$(A)\ \dfrac{5}{3}$

$(B)\ 4$

$(C)\ \dfrac{3}{5}$

$(D)\ 2$

Answer ______________

7. Let f be a function that is differentiable for all x. The value of $f'(x)$ is given for several values of x in the table below.

If $f'(x)$ is always decreasing, which of the following must be true?

I. $f(x)$ is an odd function.

II. $f(x)$ has a point of inflection at $(0, f(0))$.

III. $f(x)$ has a relative maximum at $x = 0$.

x	-12	-6	0	6	12
$f'(x)$	5	2	0	-2	-5

(A) II only (B) III only (C) II and III only (D) I, II and III

Answer____________________

8. $\int x^2 \sec^2(x^3)dx =$

$(A)\ \frac{1}{3}\tan^2(x^3) + C$

$(B)\ 3\tan(x^3) + C$

$(C)\ 3\sec(x^3) + C$

$(D)\ \frac{1}{3}\tan(x^3) + C$

Answer ____________________

9. $\int_{-\frac{\pi}{2}}^{\frac{\pi}{2}} (x - \cos 3x)dx =$

$(A)\ \frac{\pi^2}{4} + \frac{2}{3}$

$(B)\ \frac{2}{3}$

$(C)\ \frac{1}{3}$

$(D)\ 0$

Answer ____________________

10. $\int \frac{1}{x\sqrt{9-(\ln x)^2}} dx = ?$

$(A)\sin^{-1}\left(\frac{\ln x}{3}\right)+C$ $(B)\sin^{-1}\left(\frac{\ln x}{9}\right)+C$ $(C)\sec^{-1}\left(\frac{\ln x}{3}\right)+C$ $(D)\frac{1}{x}\sin^{-1}\left(\frac{\ln x}{3}\right)+C$

Answer ____________________

Examination II

11. Let $g(x)$ be the inverse of $f(x)$. The selected values of $f(x)$ and $f'(x)$ are given in the table below. Which of the following is the value of $g'(3)$?

x	0	1	3	5
$f(x)$	-2	3	1	2
$f'(x)$	7	5	8	4

(A) $\frac{1}{8}$

(B) $\frac{1}{7}$

(C) 3

(D) $\frac{1}{5}$

Answer ____________

12. The slope of function $y = f(x)$ is given by $6x^4 - 12x^2 + 2$. What are the x-coordinates of all points of inflections of $y = f(x)$?

(A) $-\frac{\sqrt{3}}{3}, \frac{\sqrt{3}}{3}$

(B) $-1, \frac{\sqrt{3}}{3}, 1$

(C) 0, 1

(D) $-1, 0, 1$

Answer ____________

13. $\int \frac{x^2-2}{x^3-6x+1}dx =$

(A) $\ln\left|x^3-6x+1\right|+C$

(B) $\frac{1}{3}\ln\left|x^3-6x+1\right|+C$

(C) $\frac{1}{3}\left(x^3-6x+1\right)+C$

(D) $\frac{\left(x^3-6x+1\right)^2}{2}+C$

Answer ____________

14. The volume of the solid formed by revolving the region bounded by the graphs of $y=x^2+2$, $y=x+1$, $x=0$, and $x=1$ about the horizontal line $y=4$ is given by which of the following ?

(A) $\pi\int_0^1\left(\left(3-x\right)^2-\left(2-x^2\right)^2\right)dx$

(B) $\pi\int_0^1\left(\left(x+1\right)^2-\left(x^2+2\right)^2\right)dx$

(C) $\pi\int_0^1\left(\left(2-x^2\right)^2-\left(3-x\right)^2\right)dx$

(D) $\pi\int_0^1\left(\left(3-x\right)^2-\left(6-x^2\right)^2\right)dx$

Answer ____________

15. Shown below is the slope field of the differential equation. What could be the solution to this differential equation?

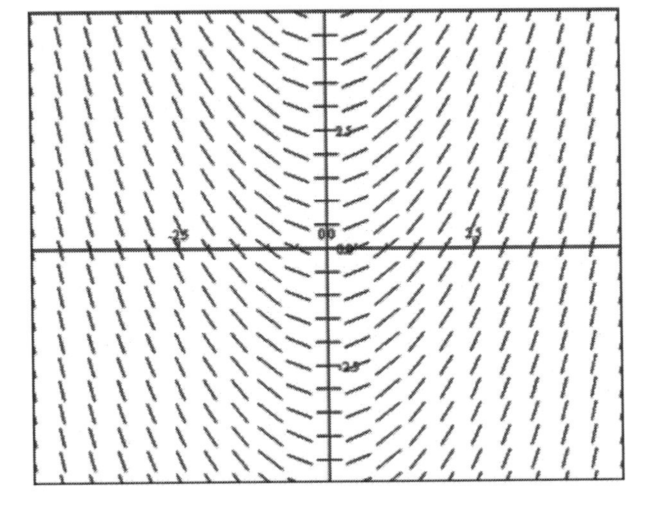

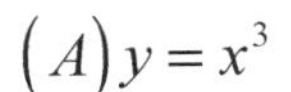

(A) $y = x^3$

(B) $y = -5x^2$

(C) $y = x$

(D) $y = x^2$

Answer ______________________

16. Which of the following is the derivative of $\sin^4(\cot^3 7x)$?

(A) $84\sin^3(\cot^3 7x)\cot^2 7x$

(B) $-84\sin^3(\cot^3 7x)\cos(\cot^3 7x)\csc^2 7x$

(C) $-84\sin^3(\cot^3 7x)\cos(\cot^3 7x)\cot^2 7x\csc^2 7x$

(D) $12\sin^3(\cot^3 7x)\cos(\cot^3 7x)\cot^2 7x\csc^2 7x$

Answer ______________________

17. The function $f(x) = x^4 + 3x^3 + 2x + 4$ must have a zero between which of the following values of x ?

$(A) -2$ and 1

(B) 1 and 2

(C) 2 and 3

(D) 3 and 4

Answer____________________

18. $\lim_{h \to 0} \dfrac{\cos\left(\dfrac{\pi}{3} + h\right) - \dfrac{1}{2}}{h} =$

$(A)\ \sqrt{3}$ $(B) -\frac{\sqrt{3}}{2}$ $(C)\ -\frac{1}{2}$ $(D)\ \frac{\sqrt{3}}{2}$

Answer ____________________

19. What are y coordinates of the points where $y^3+x^2+30y^2=4x+3$ has vertical tangent lines?

$(A)\ y=0$ $(B)\ y=-20$ $(C)\ y=0$ and $y=-20$ $(D)\ y=-20$ and $y=2$

Answer________________

20. The region is bounded by the curve of $y=\sin^2 x$ (graph of which is shown below) and the $x-$axis from $x=0$ to $x=\pi$. If this region is revolved around the x-axis, which of the following is the volume of the resulting solid?

$(A)\ \frac{3}{8}\pi$ $(B)\ \frac{1}{2}\pi^2$ $(C)\ \frac{3}{8}\pi^2$ $(D)\ \frac{3}{4}\pi^2$

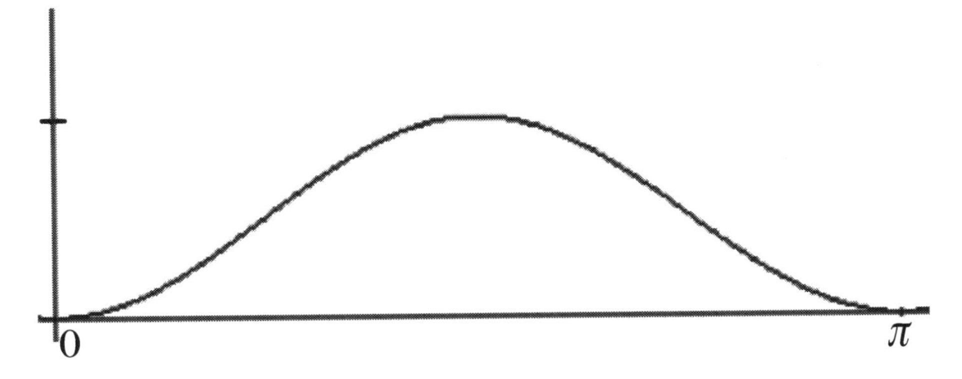

Answer________________

Examination II

21. Let $f(x)$ be a differentiable function defined on the interval $-6 \le x \le 6$. The values of $f(x)$ and its derivative $f'(x)$ at several points of the domain are shown on the table below. The equation of the line tangent to the graph of $f(x)$ and parallel to the segment between the endpoints of $f(x)$ is

$(A)\ y=-x+3$ $(B)\ y=-3x+7$ $(C)\ y=-3x+4$ $(D)\ y=-4x+9$

x	-6	-4	-2	0	2	4	6
$f(x)$	41	25	13	3	2	4	5
$f'(x)$	-5	-4	-3	-1	0	1	0

Answer____________________

22. Which of the following graphs is the graph of a differentiable function $f(x)$ that satisfies the given conditions:

$f(3)=2, f'(0)=f'(3)=f'(5)=0,$

$f'(x)<0$ if $x<0$ or $3<x<5$?

(A)

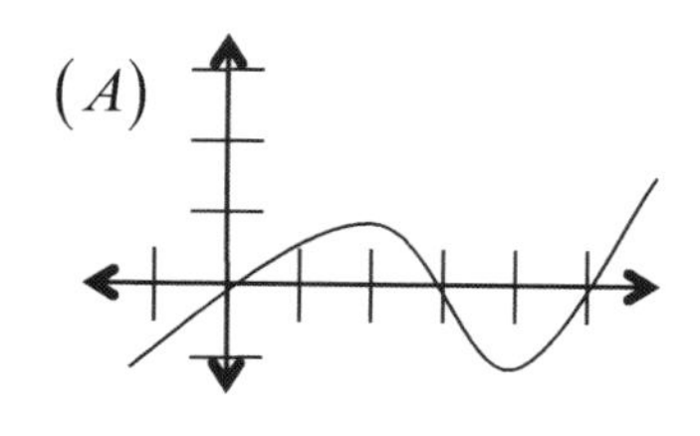

(B)

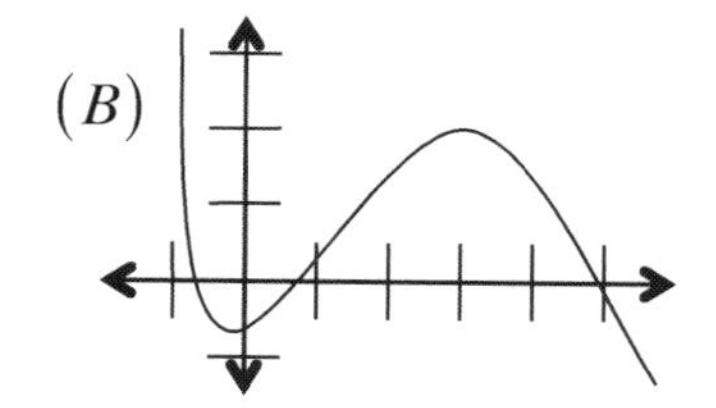

(C)

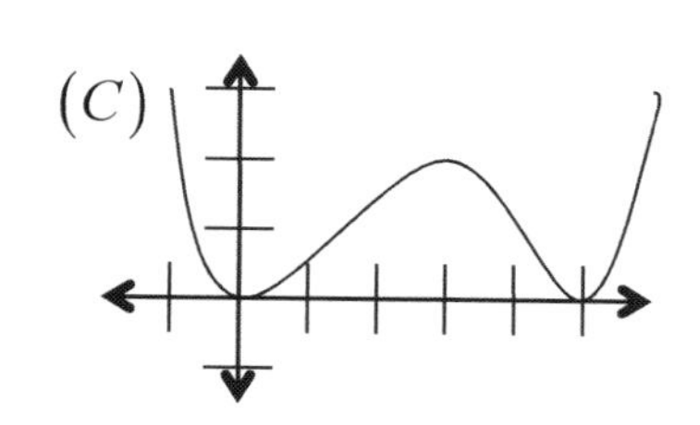

(D)

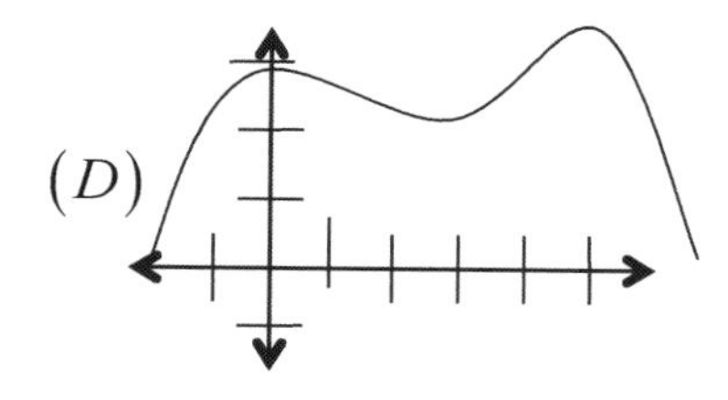

Answer ____________________

23. Given the following, which of the functions f, g, and h has at least one point of inflection?

I. $f(x) = 2\sin 2x + 2$ II. $g'(x) = 5x^3$ III. $h(x) = \frac{6}{x} - \frac{1}{2x^2}$, $x \neq 0$

(A) I only
(B) I and II only
(C) I and III only
(D) I, II, and III

Answer ____________

24. Let $\frac{dy}{dx} = e^{x-y}$. Which of the following is the solution to this equation such that $y(0) = 1$?

(A) $y = \ln x$
(B) $y = \ln(e^x + e)$
(C) $y = e^x$
(D) $y = \ln(e^x + e - 1)$

Answer ____________

25. The graph of $f'(x)$ shown has horizontal tangents at $(4,0)$ and $(12,-6)$. What are the $x-$coordinates for the point(s) of inflection of function $f(x)$?

(A) 4 only (B) 4 and 15 only (C) -3 and 20 only (D) 4 and 12 only

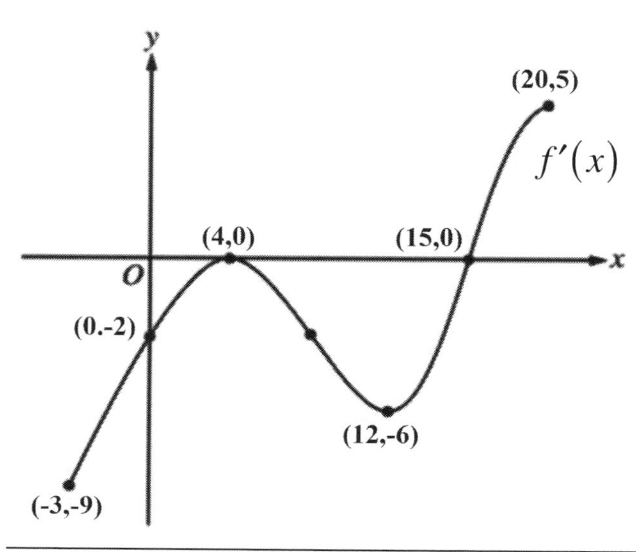

Answer ______________

26. A wire 80 inches long is to be cut into two pieces. One piece is to be bent into the shape of a circle and the other into a square. What should be the length of the smaller piece so that the sum of the areas of the circle and the square is minimized?

(A) $\frac{80\pi}{8+2\pi}$ (B) $\frac{80\pi}{4+\pi}$ (C) $\frac{160}{4+\pi}$ (D) $\frac{80}{4+\pi}$

Answer ______________

27. If n is a positive integer, then $\lim_{n\to\infty}\frac{1}{n}\left[\cos\frac{\pi}{n}+\cos\frac{2\pi}{n}+...+\cos\frac{n\pi}{n}\right]$ can be expressed as

$(A)\int_0^1 \cos\frac{\pi}{n}dx$ $(B)\pi\int_0^1 \cos x dx$ $(C)\int_0^1 \cos(\pi x)dx$ $(D)\int_0^{\pi} \cos \pi x dx$?

Answer________________________

28. Given the function $f(x)=3x^3+ax^2+bx+k$ where a, b, and k are constants. This function has a local minimum at $x=-2$ and a point of inflection at $x=-3$. What is the value of b?

(A) 81

(B) 72

(C) 54

(D) 27

Answer ________________________

29. The acceleration of a particle moving along a horizontal line at time t is $2\tan t\sec^2 t$. If the initial velocity is 1 and the initial position is 0,what is the position of the particle at time $t=\frac{\pi}{4}$?

$(A)-1$ $(B)1$ $(C)0$ $(D)2$

Answer________________

30. A solid has a circular base of radius 4. If every plane cross section perpendicular to the x-axis is an equilateral triangle, then the volume is

$(A)\frac{256}{3}$

$(B)\frac{64}{3}\sqrt{3}$

$(C)\frac{128}{3}\sqrt{3}$

$(D)\frac{256}{3}\sqrt{3}$

Answer________________

Section I Part B

Directions: Solve each of the following problems, using the space provided. Choose the best answer. Do not spend too much time on any one problem. A graphing calculator is required for some questions on this part of the exam.

In this test:

1. The exact numerical value of the correct answer does not always appear among the choices given. Then select from among the choices the number that best approximates the exact numerical value.
2. Unless otherwise specified, the domain of a function is assumed to be the set of all real numbers x for which $f(x)$ is a real number.

31. If f is an antiderivative of $\frac{\sin^2 x}{x^2+2}$ such that $f(2)=\frac{1}{2}$, then $f(0)=?$

(A) -0.325

(B) 0.175

(C) 0.825

(D) 1.175

Answer ____________________

32. The graph of the function $f(x)=4\cos x$ crosses the y-axis at the point $P(0,4)$ and the x-axis at the point $Q\left(\frac{\pi}{2},0\right)$. Which of the following is the x-coordinate of the point on the graph of $f(x)$, between P and Q at which the line tangent to the graph of $f(x)$ is parallel to the segment connecting points P and Q?

(A) 0.6901

(B) 0.8901

(C) 1.0601

(D) 0.1348

Answer ____________________

33. The figure below shows the graph of f', the derivative of the function f, on the closed interval $-1 \le x \le 5$. The graph of f' has horizontal tangent lines at $x = 1$ and $x = 3$. The function f is twice differentiable with $f(3) = 8$. Let g be the function defined by $g(x) = x^2 f(x)$. Which of the following is the equation for the line tangent to the graph of g at $x = 3$?

(A) $y - 24 = 72(x + 3)$

(B) $y - 9 = 21(x - 3)$

(C) $y - 72 = 21(x - 3)$

(D) $y - 36 = 14(x - 3)$

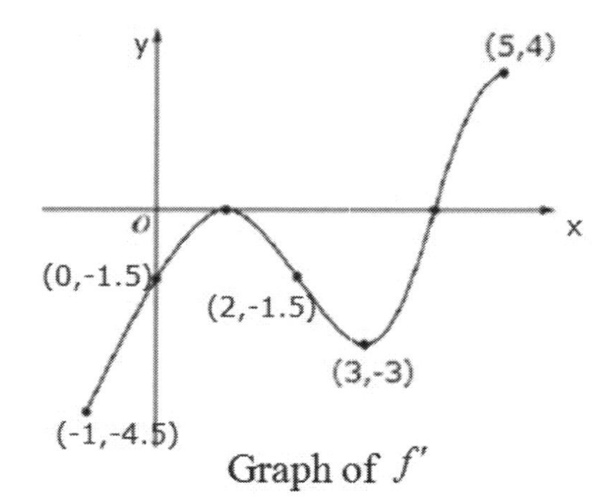

Graph of f'

Answer ______________________

34. The position $s(t)$ of a point P moving along a line is given by the function $s(t) = 5t^3 - 2t^2 + 8$ with t in seconds and $s(t)$ in centimeters. What is the average velocity in cm/sec of point P on the time interval $[1, 1.4]$?

(A) 32.00

(B) 34.93

(C) 17.00

(D) 16.00

Answer ______________________

35. Let f be the function defined by $f(x)=\begin{cases}\sqrt{x+2} & \text{for } 0\le x\le 2\\ 4-x & \text{for } 2<x\le 5.5\end{cases}$

What is the average value of $f(x)$ on the closed interval $0\le x\le 5.5$?

(A) 0.786 (B) 0.891 (C) 1.043 (D) 0.915

Answer____________________

36. A ball is thrown up into the air at the velocity of 50 ft/sec. After t seconds, the distance of the ball above the ground is $50t-16t^2$ ft. What is the velocity of the ball the instant it strikes the ground?

(A) 25 ft/sec

(B) -25 ft/sec

(C) 50 ft/sec

(D) -50 ft/sec

Answer ____________________

37. A sphere is expanding in such a way that the area of any circular cross section through the sphere's center is increasing at a constant rate of 3 $cm^2 / \sec$. At the instant when the radius of the sphere is 6 centimeters, what is the rate of change of the sphere's volume? (The volume V of a sphere with radius r is given by $V = \frac{4}{3}\pi r^3$.)

(A) $396\pi\ cm^3/\text{sec}$

(B) $72\ cm^3/\text{sec}$

(C) $36\ cm^3/\text{sec}$

(D) $9\pi\ cm^3/\text{sec}$

Answer____________________

38. A graph of a function consists of a line segment from the point $(0,10)$ to point $(8,8)$, another line segment from $(8,8)$ to $(12,8)$, and a third line segment from $(12,8)$ to $(20,0)$. What is the average value of this function on the interval $[0,20]$?

(A) 136

(B) 7

(C) $\frac{34}{5}$

(D) $\frac{127}{20}$

Answer ____________________

39. A reactor coolant tank is being filled with coolant at the rate of $400\sqrt[3]{t}$ gallons per hour with $t > 0$ measured in hours. If the tank originally contained 200 gallons of coolant, how many gallons are in the tank after 8 hours?

(A) 800

(B) 4800

(C) 5000

(D) 6400

Answer ______________________

40. Let P be the partition of an interval $[2,6]$ determined by $\{2,4,5,6\}$. Using the Midpoint Riemann Sum, which of the following is the best approximation for the area under the curve of $f(x) = x^2 + 4$ on the interval $[2,6]$?

$(A)65$ $(B)75.25$ $(C)55$ $(D)84.5$

Answer ______________________

41. A particle, initially at rest, moves along the $x-$axis so that its acceleration at any time is given by $a(t)=8t^2-6$. Which of the following is the total distance traveled by the particle within the time interval $2 \le t \le 4$?

(A) $\frac{412}{3}$

(B) 148

(C) $\frac{340}{3}$

(D) 124

Answer________________

42. On the diagram below, the cross section of a water tank that is being drained is an equilateral triangle. The area of the cross section of water, $A(t)$, is decreasing at a rate of 10cm^2/min. Which of the following is the rate (in cm/min) at which length of its side s is changing when area of the triangle is 250 cm^2?

(A) -0.240 (B) -0.481 (C) -0.832 (D) -0.951

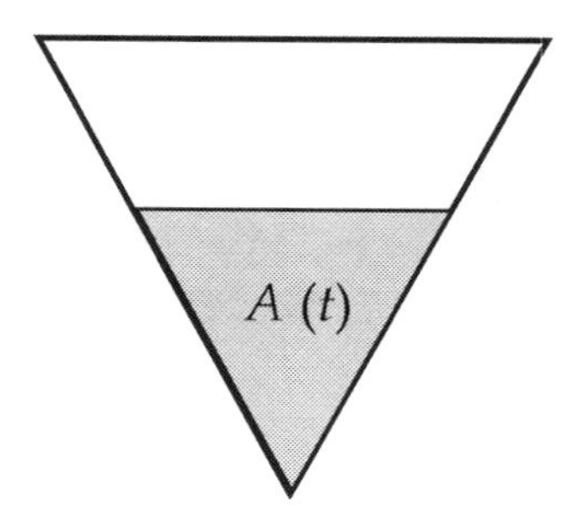

Answer________________

43. While on a road trip, a girl drives for 8 hours. Her velocity, recorded at random times in miles per hour, is given in the table below. Based on the data in the table, how many miles has she driven after 8 hours?

(A) 320 miles (B) 325 miles (C) 328 miles (D) 330 miles

t	0	1	5	6	8
$v(t)$	25	30	60	45	20

Answer ____________________

44. A pizza, heated to a temperature of 475 degrees Fahrenheit $(°\text{F})$, is taken out of an oven and placed in a 105°F room at $t = 0$ minutes. The temperature of the pizza is changing at a rate of $-256e^{-0.7t}$ degrees Fahrenheit per minute. To the nearest degree, what is the temperature of the pizza at $t = 9$ minutes?

(A) 100 (B) 110 (C) 115 (D) 120

Answer ____________________

45. The region bounded by the graphs of the equations $\left(\frac{x}{2}\right)^2 - y = -2$ and $y = \frac{x}{5} + 4$ and by the vertical lines $x = 1$ and $x = 3$ is revolved around the x-axis. What is the volume of the resulting solid?

(A) 38.110

(B) 45.312

(C) 59.863

(D) 64.822

Answer________________

Examination III

Section I Part A

Directions: Solve each of the following problems, using the space provided. Choose the best answer. Do not spend too much time on any one problem. Calculators may NOT be used on this part of the exam.

In this test: Unless otherwise specified, the domain of a function is assumed to be the set of all real numbers x for which $f(x)$ is a real number.

1. Let f be defined as follows,

$$f(x)=\begin{cases} x^2+1, & \text{for } -2\le x<2 \\ 1, & \text{for } x=2 \\ 5, & \text{for } 2<x<\infty \end{cases}$$

What is the limit of $f(x)$ as x approaches 2?

(B) 1 (C) 2 (C) 5 (D) nonexistant

Answer________________________

2. What is the number represented by the difference between the average rate of change of function $y=f(x)$ with respect to x on the given interval $[1,1.5]$ and the instantaneous rate of change of y with respect to x at the right endpoint of this interval if $y=4x^2-1$?

(A)1

(B)2

(C)6

(D)8

Answer ________________________

3. $\lim_{h\to 0} \frac{\cot(3(x+h)) - \cot(3x)}{h} =$

$(A) -\csc^2 3x$

$(B) -3\csc 3x \cot 3x$

$(C) \csc 3x \cot 3x$

$(D) -3\csc^2 3x$

Answer____________________

4. Given $xy^2 = 4$, what is the value of the second derivative of the function $y = f(x)$ at the point $(1,2)$?

$(A)\ \frac{3}{2}$

$(B)\ \frac{3}{4}$

$(C)\ 1$

$(D)\ \frac{1}{2}$

Answer ____________________

5. What is y if $\frac{dy}{dx} = \frac{3x+2}{5y}$ and $y = 1$ when $x = 2$?

$(A)\, y = \sqrt{\frac{3x^2+4x-15}{5}}$ $(B)\, y = \pm\sqrt{\frac{3x^2+4x-15}{5}}$

$(C)\, y = \pm\sqrt{\frac{3x^2+4x+13}{2}}$ $(D)\, y = \sqrt{\frac{3x^2+4x}{5}}$

Answer____________________

6. $\left(\tan^{-1}\left(e^{x^2}\right)\right)' = ?$

$(A)\frac{2xe^{x^2}}{1+e^{2x^2}}$ $(B)\frac{4xe^{x^2}}{1+e^{x^4}}$ $(C)\frac{2x}{1+e^{2x^2}}$ $(D)\frac{2xe^{x^2}}{1+e^{x^2}}$

Answer ____________________

7. $\lim_{h \to 0} \frac{1}{\ln(h+1)} \int_0^h \frac{\sin t}{t^2+t} dt$ is

$(A)\,0$ $(B)\,1$ $(C)\,\pi$ (E) nonexistent

Answer________________

8. Which of the following is the maximum area of the rectangle that can be inscribed in a semicircle of radius $4\sqrt{2}$, if two vertices lie on the diameter?

(A) 24

(B) 32

(C) 36

(D) 64

Answer ________________

9. $\frac{d}{dx}\int_0^{x^3}\sqrt{t^2+2}\,dt=$

$(A)\ \sqrt{x^6\cdot 3x^2+2}-\sqrt{2}$

$(B)\ \sqrt{x^6+2}$

$(C)\ \left(\sqrt{x^6+2}\right)\cdot 3x^2$

$(D)\ \left(\sqrt{x^6+2}\right)\cdot x^3$

Answer____________________

10. $\int_4^6 |x-5|\,dx=$

$(A)4$

$(B)0$

$(C)1$

$(D)2$

Answer ____________________

11. What is $f'(x)$ if $f(x)=(3x^2+1)\cdot 5^{\frac{1}{x^2}}$?

$(A)\ 6x\cdot 5^{x^{-2}}+\left(5^{x^{-2}}\cdot\ln 5\right)\left(-2x^{-3}\right)$

$(B)\ 6x\cdot 5^{x^{-2}}+\left(3x^2+1\right)\left(5^{x^{-2}}\cdot\ln 5\right)$

$(C)\ 6x\cdot 5^{x^{-2}}+\left(3x^2+1\right)\left(5^{x^{-2}}\cdot\ln 5\right)\left(-2x^{-3}\right)$

$(D)\ 3x^2\cdot 5^{x^{-2}}+\left(3x^2+1\right)\left(5^{x^{-2}}\cdot\ln 5\right)\left(-2x^{-3}\right)$

Answer____________________

12. The graph of $f(x)$ is shown below. List the following values of $f'(1), f'(-2), f''(0)$ from smallest to largest.

$(A)\ f'(-2),\ f''(0),\ f'(1)$

$(B)\ f''(0),\ f'(1),\ f'(-2)$

$(C)\ f''(0),\ f'(-2),\ f'(1)$

$(D)\ f'(1),\ f''(0),\ f'(-2)$

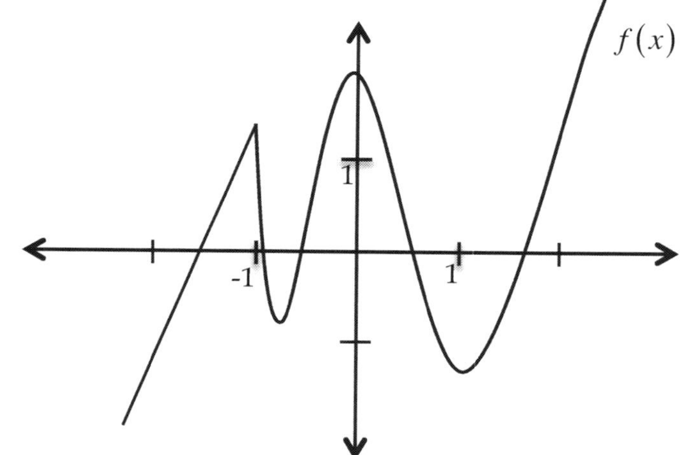

Answer ____________________

13. $\int \frac{dx}{1+\sin x} =$

(A) $-\frac{1}{(1+\sin x)^2}+C$

(B) $\tan x+\sec x+C$

(C) $\ln|1+\sin x|+C$

(D) $\tan x-\sec x+C$

Answer_______________

14. The function $h(x)=\frac{f(x)}{g(x)}$ where $f(x)$ and $g(x)$ are piecewise linear functions whose graphs are shown below. What is $h'(-2)$?

$(A) -\frac{3}{2}$ $(B) -1$ $(C) -\frac{1}{2}$ $(D) 1$

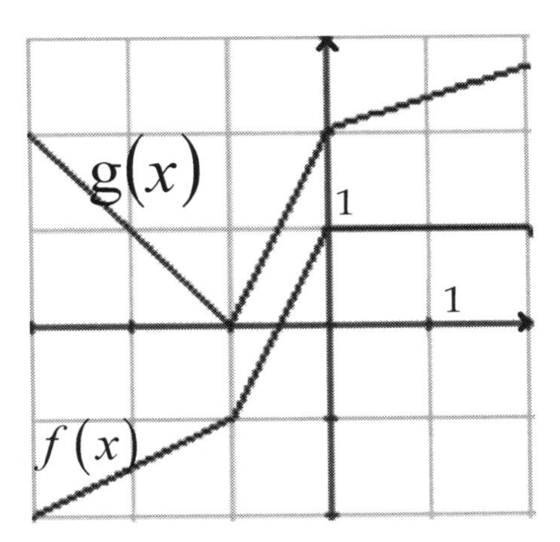

Answer _______________

15. The graph of $f(x)$ is the semicircle with radius 5 as shown below. What is the mean value of $f(x)$ on the interval $[0,10]$?

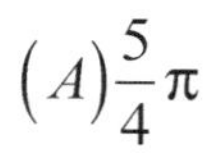

$(A)\ \frac{5}{4}\pi$

$(B)\ \frac{5}{2}\pi$

$(C)\ 5\pi$

$(D)\ \frac{25\pi}{2}$

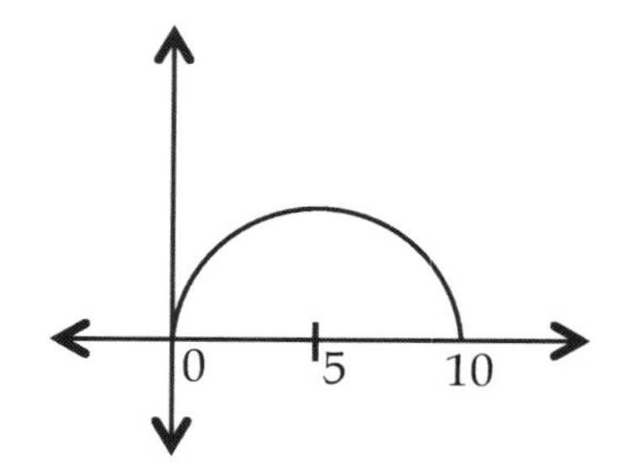

Answer____________________

16. Given f is continuous for all real numbers, $\frac{dy}{dx} = f(x)$ and $y(-1) = 6$, what is $y(x)$?

$(A)\ 6 + \int_{-1}^{x} f(t)dt$

$(B)\ \int_{-1}^{x} f(t)dt - 6$

$(C)\ 6 + \int_{-1}^{x} f'(t)dt$

$(D)\ \int_{-1}^{x} f'(t)dt - 6$

Answer ____________________

17. $\frac{d}{dx}\int_{4x}^{x^4}\left(t^4+2\right)^{11}dt=$

$(A)\ 4x^3\left(\left(x^4\right)^4+2\right)^{11}-4\left(\left(4x\right)^4+2\right)^{11}$

$(B)\ \left(\left(x^4\right)^4+2\right)^{11}-\left(\left(4x\right)^4+2\right)^{11}$

$(C)\ 4x^3\left(\left(x^4\right)^4+2\right)^{11}$

$(D)\ x^4\left(\left(x^4\right)^4+2\right)^{11}-4x\left(\left(4x\right)^4+2\right)^{11}$

Answer__________________________

18. What is the area of the region bounded by the curve $x = y^2 - 4$ and the y-axis?

$(A)\frac{28}{3}$

$(B)10$

$(C)\frac{32}{3}$

$(D)\frac{34}{3}$

Answer __________________________

19. A particle moves horizontally with the velocity $v=\cos^2 \pi t$ from time $t=0$ to $t=10$. What is the total distance the particle travels in this time?

$(A)2 \quad (B)\frac{5}{2} \quad (C)4 \quad (D)5$

Answer____________________

20. What is the derivative of function $y(x)$ if $\tan(2y)=xe^y$?

$(A)\frac{e^y}{2\sec^2(2y)-xe^y} \quad (B)\frac{e^y+xe^y}{2\sec^2(2y)} \quad (C)\frac{e^y}{\sec^2(2y)-xe^y} \quad (D)\frac{e^y}{2\sec(2y)\tan(2y)-xe^y}$

Answer ____________________

21. If $f(x) = \log \sqrt[4]{(3x+5)^3}$, where $x > -\frac{5}{3}$, then $f'(x) =$

$(A)\ \frac{3}{4(3x+5)\ln 10}$

$(B) -\frac{3}{4(3x+5)\ln 10}$

$(C)\ \frac{9}{4(3x+5)\ln 10}$

$(D)\ \frac{9}{(3x+5)\ln 10}$

Answer____________________

22. $\int \frac{\cos 2x}{\sin^2 x \cdot \cos^2 x} dx =$

$(A) - \tan x + \cot x + C$

$(B) - \tan x - \cot x + C$

$(C) \tan x - \cot x + C$

$(D) \ln|\csc x - \cot x| -$
$\ln|\sec x + \tan x| + C$

Answer ____________________

23. Function $f(x)$ is a linear function such that $f(x+2)-f(x)=8$. Given that $g(x)$ is the inverse of $f(x)$, which of the following is $g'(x)$?

$(A)\ \frac{1}{8}$

$(B)\ \frac{1}{4}$

$(C)\ \frac{1}{2}$

$(D)\ 4$

Answer________________

24. What is $f'(x)$ if $f(x)=4^{\arctan(2x^5)}$?

$(A)\ 4^{\arctan(2x^5)}\cdot\sec^2(2x^5)\cdot\ln 4\cdot 10x^4$

$(B)\ 4^{\arctan(2x^5)}\cdot\frac{1}{\left(1+(2x^5)^2\right)\ln 4}\cdot 10x^4$

$(C)\ 4^{\arctan(2x^5)-1}\cdot\frac{10x^4}{1+(2x^5)^2}$

$(D)\ 4^{\arctan(2x^5)}\cdot\ln 4\cdot\frac{1}{1+(2x^5)^2}\cdot 10x^4$

Answer ________________

25. The velocity of a particle moving along a straight line is given by $v(t) = 4\pi \sin \pi t$. How far does the particle travel between $t = 0$ and $t = 2$?

(A) 32 (B) 16 (C) 8 (D) 0

Answer________________

26. What are all the horizontal asymptotes for the graph of $f(x) = \dfrac{5x}{\sqrt{x^2 + 1}}$?

(A) Horizontal asymptote is $y = 0$

(B) Horizontal asymptote is $y = 5$

(C) Horizontal asymptote is $y = -5$

(D) Horizontal asymptotes are $y = 5$ and $y = -5$

Answer ________________

27. If n is a positive integer, then $\lim_{n\to\infty}\frac{1}{n^3}\left[1^2+2^2+\ldots+n^2\right]$ can be expressed as

$(A)\int_0^1 x^2\,dx$ $(B)\int_0^1 \frac{1}{x}\,dx$ $(C)\int_0^1 \frac{1}{x^3}\,dx$ $(D)\int_0^1 x^3\,dx$

Answer____________________

28. Shown on the right is the slope field for which differential equation?

$(A)\ \frac{dy}{dx}=\frac{y^2-2y}{x}$

$(B)\ \frac{dy}{dx}=\frac{3x^2-4}{x}$

$(C)\ \frac{dy}{dx}=\frac{3y^2-4}{4}$

$(D)\ \frac{dy}{dx}=\frac{y-2}{2x}$

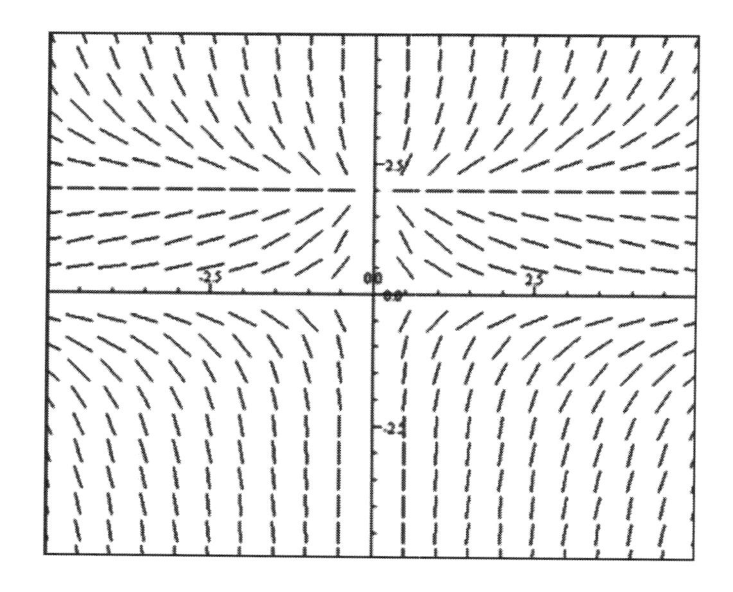

Answer ____________________

29. If $\frac{dy}{dx} = y^2(8-2x)$ and $x = 2$ when $y = \frac{1}{3}$, then which of the following is equal to y?

$(A)\frac{1}{x^2 - 8x + 12}$

$(B)\frac{1}{x^2 - 8x + 15}$

$(C)\frac{1}{-x^2 + 8x - 9}$

$(D)\frac{1}{x^2 - 8x}$

Answer________________________

30. Let function $f(x)$ be positive while its first derivative $f'(x)$, and its second derivative $f''(x)$ all be negative on a closed interval $[a,b]$. The interval $[a,b]$ is partitioned into n equal length sub-intervals and these are used to compute an Upper Sum U, a lower sum L, and the Trapezoidal Rule Approximation T. If $I = \int_a^b f(x)dx$, which statement below is true?

(A) L < U < T < I

(B) L < T < I < U

(C) L < T < U < I

(D) L < I < T < U

Answer ________________________

Examination III

Section I Part B

Directions : Solve each of the following problems, using the space provided. Choose the best answer. Do not spend too much time on any one problem. A graphing calculator is required for some questions on this part of the exam.

In this test :

1. The exact numerical value of the correct answer does not always appear among the choices given. Then select from among the choices the number that best approximates the exact numerical value.
2. Unless otherwise specified, the domain of a function is assumed to be the set of all real numbers x for which $f(x)$ is a real number.

31. A particle moves along the $x-$axis so that at any time $t > 0$, its acceleration is given by $a(t) = \sin t - 3\sin t\cos t + 2t$. If the velocity of the particle is 6 m / sec at time $t = 3$ sec, then the velocity in m / sec of the particle at time $t = 7$ sec is approximately equal to

(A)43.639 (B)31.640 (C)37.639 (D)39.366

Answer____________________

32. As shown in the figure below the function $f(x)$ consists of a line segment from $(0,0)$ to $(3,5)$, another line segment from $(3,5)$ to $(7,5)$, and a quarter-circle with radius of 5. What approximately is the average value of this function on the interval $[0,\ 12]$?

(A) 3.470 (B) 3.511 (C) 3.928 (D) 41.635

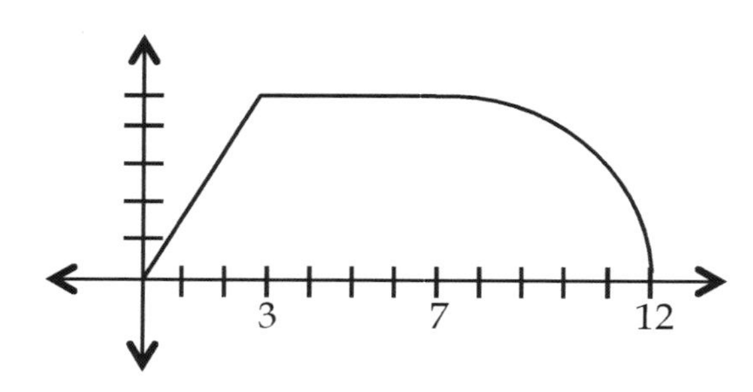

Answer ____________________

33. The figure below shows the graph of the derivative of a function f. How many points of inflection does f have in the interval shown on the diagram?

(A) Zero

(B) One

(C) Two

(D) More than two

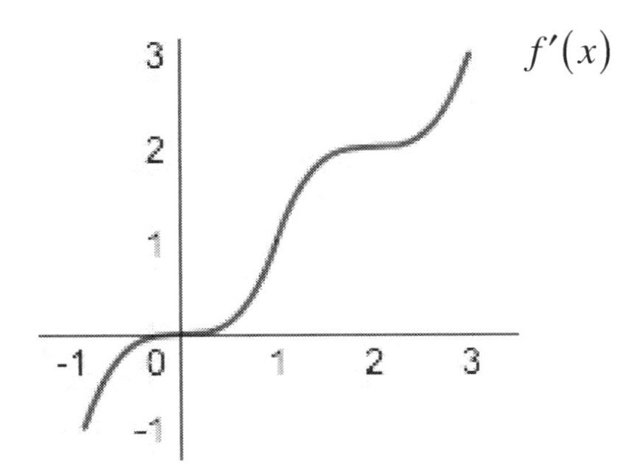

Answer________________

34. An object moves along the $x-$axis so that its position at any time t is given by $x(t) = -4\cos\frac{t}{2} + 3$. For which value of t is its speed the greatest on the interval $[4,8]$?

(A) $t = 4$

(B) $t = 5$

(C) $t = 6$

(D) $t = 7$

Answer________________

35. If f is a continuous function defined by

$$f(x)=\begin{cases} x^2-bx, & x\le 3 \\ 3\cos\left(\frac{\pi}{4}x\right), & x>3 \end{cases}$$

then $b\approx$

$(A)-2.2929$ $(B)\,3.7071$ $(C)-3.7071$ $(D)\,2.2929$

Answer________________________

36. A balloonist drops a sandbag from a balloon 160 feet above the ground. After t seconds, the sandbag is $160-16t^2$ feet above the ground. With what velocity does the sandbag strike the ground?

$(A)\ -32\sqrt{10}$

$(B)\ -16\sqrt{10}$

$(C)\ -24\sqrt{10}$

$(D)\ -24$

Answer ________________________

37. An outdoor thermometer registers a temperature of $100^\circ F$. Twelve minutes after it is brought into a room where the temperature is $80^\circ F$, the thermometer registers $90^\circ F$. When will the thermometer register $85^\circ F$?
(A) in 20 minutes (B) in 24 minutes (C) in 32 minutes (D) in 48 minutes

Answer______________________

38. A particle moves along the x-axis such that its acceleration from time $t = 0$ to time $t = 5$ is given by $a(t) = -2\sin(t)$. The particle has a velocity of 1 at time $t = 0$. What is the total distance traveled by the particle over the time interval?
(A) 6.918 (B) 8.288 (C) 7.640 (D) 3.288

Answer ______________________

39. The region bounded by the graphs of the equations $(x-4)^2 = y-2$ and $2y-3x+1=0$ is revolved around the x-axis. The volume of the solid generated is

(A) 162.90

(B) 218.36

(C) 342.99

(D) 436.71

Answer______________________

40. A solid has its base the region bounded by the graph of $y = \dfrac{10}{1+x^2}$ and the horizontal line $y=3$. What is the volume of the solid if every cross section by a plane perpendicular to the x-axis is a semicircle?

(A)21.009 (B)20.672 (C)17.442 (D)10.374

Answer ______________________

41. A boy is standing on a dock watching a boat moving north away from him, at a speed of $5000 \text{ft} / \text{min}$. A girl is standing 1000 ft to the east of the boy and is watching the same boat. How fast is the boat moving away from the girl when it is 12500 ft away from the boy?

$(A)538.280$ $(B)5023.949$ $(C)4984.08$ $(D)4321.957$

Answer________________________

42. The region bounded by the graph of $y = -x^2 + 2$ and the $x-$axis is revolved around $y = 3$. What is the volume of the resulting solid?

$(A)52.130$ $(B)12.506$ $(C)24.600$ $(D)27.842$

Answer ________________________

43. The number of a certain bacteria doubles every 15 sec. Its rate of growth is directly proportional to the number of bacteria present. If the initial number of bacteria is 75,000, what is the number of bacteria after 15 min?

(A) 6.485×10^{27} (B) 9.067×10^{28} (C) 8.647×10^{22} (D) 6.452×10^{15}

Answer________________________

44. Let $f(x)$ be a function that is differentiable for all x. The table below gives the value of $f'(x)$ for several values of x. If $f'(x)$ is always increasing, which statement about $f(x)$ must be true?

(A) $f(x)$ is an odd function.

(B) $f(x)$ is increasing on $[-4,4]$.

(C) $f(x)$ has a relative minimum at $x=0$

(D) $f(x)$ has a relative maximum at $x=0$.

x	-4	-2	0	2	4
$f'(x)$	-3	-1	0	1	3

Answer ________________________

45. A rectangle with vertices $(0,0)$, $(8,0)$, $(8,10)$ and $(0,10)$ is shown below. The graph of $y = x + \sqrt[3]{x}$ divides it into two regions. What is the probability that a point picked randomly from inside of this rectangle lies in the region above the graph?

$(A)\ \frac{15}{32}$ $(B)\ \frac{9}{20}$ $(C)\ \frac{7}{16}$ $(D)\ \frac{3}{8}$

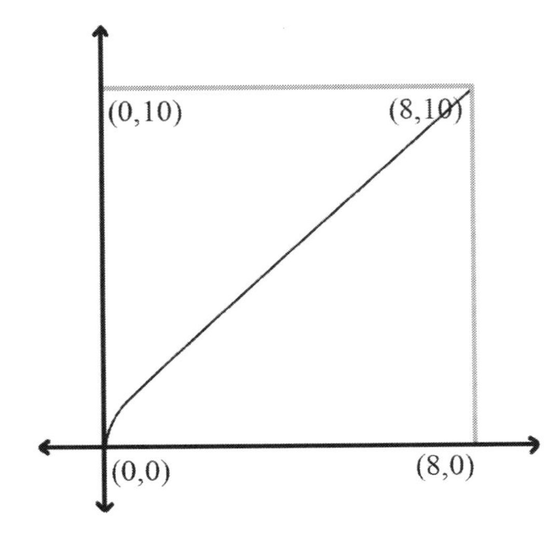

Answer____________________

Examination IV

Section I Part A

Directions : Solve each of the following problems, using the space provided. Choose the best answer. Do not spend too much time on any one problem. Calculators may NOT be used on this part of the exam.

In this test : Unless otherwise specified, the domain of a function is assumed to be the set of all real numbers x for which $f(x)$ is a real number.

1. $\lim_{x\to 25} \frac{x-25}{\sqrt{x}-5} =$

(A) 5

(B) 10

(C) 15

(D) 20

Answer ________________

2. Let $f(x)$ be a continuous function. What is the value of $\int_3^{15} f(x)dx$ if it is given that $\int_1^5 f(3x)dx = 7$?

$(A)\ \frac{7}{3}$ (B) 7 (C) 14 (D) 21

Answer ________________

3. If $f(x)=e^e$, which of the following is $f'(x)$?

$(A)e^e$

$(B)0$

$(C)e$

$(D)e^{e-1}$

Answer ____________________

4. $\lim_{n\to 5}\dfrac{\int_{25}^{n^2}(x-5)\,dx}{10n-50}$ is

$(A)\ 5$ $(B)\ 10$ $(C)\ 20$ (D) nonexistent

Answer ____________________

5. A spherical balloon is being filled with air. The radius increases at a rate of 1 inch per second. What is the instantaneous rate of change in volume when the radius of the balloon is 4 inches?

(A) 64π

(B) 36π

(C) 16π

(D) 4π

Answer________________

6. The derivative of a function $f(x)$ is continuous on $x > 0$, and is given by $f'(x) = \sin(\pi x) + \frac{7}{x^2} + 3x^2$. If the function $f(x)$ contains the point $(1,7)$, what is the y-coordinate of $f(x)$ at $x = 7$?

$(A) 348$ $(B) 348 + \frac{2}{\pi}$ $(C) 355$ $(D) 355 + \frac{2}{\pi}$

Answer________________

7. The function g is defined by $g(x)=(f(x))^3$. The graph of f' is shown below, and the point (6,12) is on the graph of f'. If $f(6)=-5$, which of the following is the slopeof the line tangent to the graph of g at $x=6$?

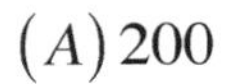

(A) 200

(B) 75

(C) 900

(D) 120

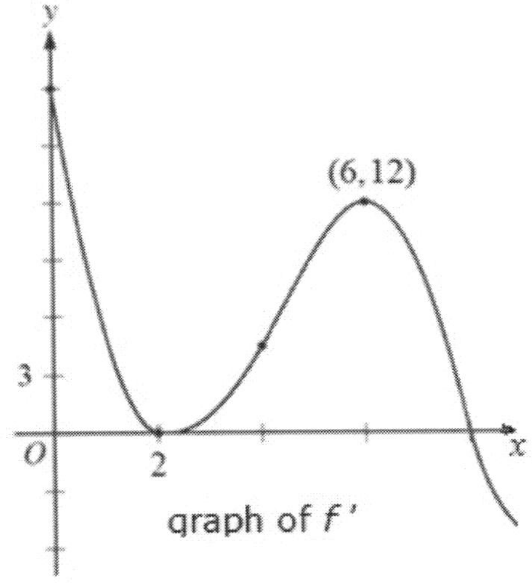

Answer ________________

8. The graph of $f'(x)$ on the interval $[-1,4]$ is shown below. On what intervals is the graph of $f(x)$ concave up?

(A) $(0.5,3.5)$

(B) $(-1,0.5)\cup(3.5,4)$

(C) $(2,4)$

(D) $(0.5,4)$

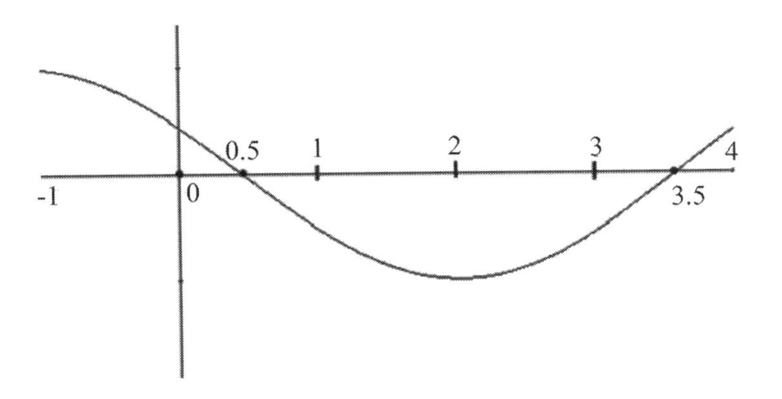

Answer ________________

9. 2000 feet of wire will be used to construct six identical rectangular enclosures as shown on the diagram below. What are the dimensions (in feet) that maximize the enclosed area of the large rectangle ?

$(A)\ l = 400, w = 200$

$(B)\ l = 500, w = 125$

$(C)\ l = \frac{1000}{3}, w = 250$

$(D)\ l = \frac{500}{3}, w = 375$

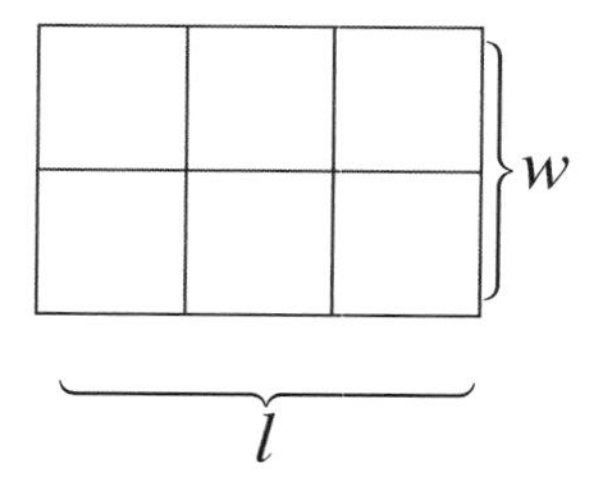

Answer________________________

10. $\int_{m}^{m+h} f(x)\ dx - \int_{m}^{h} f(x)\ dx =$

$(A) \int_{0}^{h} f(x)\ dx$

$(B) \int_{h}^{h+m} f(x)\ dx$

$(C) \int_{h}^{m} f(x)\ dx$

$(D) \int_{m}^{m+h} f(x)\ dx$

Answer________________________

11. An equation of the line tangent to the curve $x^2 + y^2 = 25$ at the point $(-4, 3)$ is

(A) $3y - 4x = 25$ (B) $4y - 3x = 25$ (C) $3y + 4x = -25$ (D) $3y - 4x = 20$

Answer ______________________

12. What is the area of the region bounded by the graphs of $y = x^2 - 4$ and $y = 3x$?

(A) $\frac{56}{3}$

(B) $\frac{45}{6}$

(C) 21

(D) $\frac{125}{6}$

Answer ______________________

13. $\int_{-1}^{4} |4-2x| \, dx =$

(A) 5

(B) 10

(C) 13

(D) 14

Answer________________________

14. The region in the first quadrant bounded by the graphs of the equations $y = x^2$ and $y = 2x^2 - 4$ is revolved around the $y-$axis. What is the volume of the solid of revolution?

(A) 2π (B) 3π (C) 4π (D) 5π

Answer ________________________

15. Shown below is the graph of $f(x)$. Which of the following limits exist?

I. $\lim_{x \to 0} f(x)$

II. $\lim_{x \to 1} f(x)$

III. $\lim_{x \to 2} f(x)$

(A) I only (B) I and II only (C) I and III only (D) I, II, and III

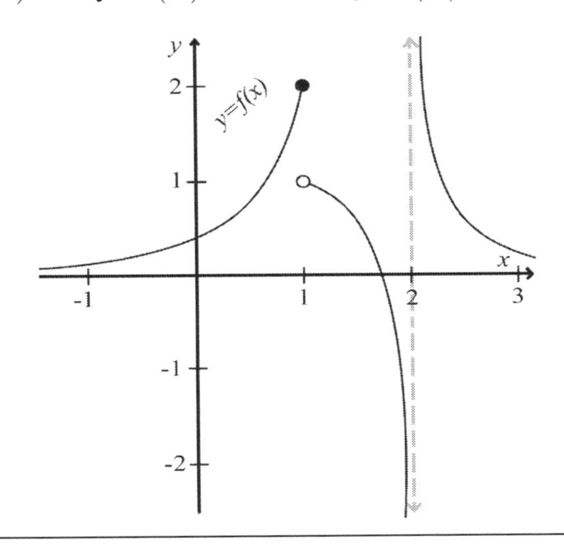

Answer ____________

t	0	1	2	3	4	5
$f(t)$	0	50	90	200	275	350

16. The table above shows the total distance $f(t)$ a bicyclist has traveled along a straight road in meters during t seconds. What is approximately the velocity of the bicyclist at $t = 4$ seconds?

(A) 75 m/sec (B) 50 m/sec (C) 40 m/sec (D) 25 m/sec

Answer ____________

17. What is the derivative of $\frac{4x^2-3x+7}{5x}$?

$(A)\ \frac{4x^2+7}{25x^2}$

$(B)\ \frac{4x^2-7}{5x^2}$

$(C)\ \frac{7-4x^2}{5x^2}$

$(D)\ \frac{8x-3}{5}$

Answer____________________

18. Which of the following is an equation of the normal line to the graph of $y=\frac{1-\cos x}{1+\cos x}$ at the point $P\left(\frac{\pi}{2}, 1\right)$?

$(A)\ y-1=-\frac{1}{2}\left(x-\frac{\pi}{2}\right)$

$(B)\ y-1=2\left(x-\frac{\pi}{2}\right)$

$(C)\ y-1=\frac{1}{2}\left(x-\frac{\pi}{2}\right)$

$(D)\ y-1=-2\left(x-\frac{\pi}{2}\right)$

Answer ____________________

19. If $f(x)=\sqrt{6\sin x+9}$, then the derivative of f at $x=0$ is

$(A)\frac{1}{2\sqrt{3}}$ $(B)0$ $(C)1$ $(D)\sqrt{3}$

Answer ____________

20. $\lim_{h\to 0}\frac{4\cos^4(x+h)+3\sin(x+h)-4\cos^4 x-3\sin x}{h}$ is

$(A)0$

$(B)16\cos^3 x\sin x+3\cos x$

$(C)-16\cos^3 x+3\cos x$

$(D)-16\cos^3 x\sin x+3\cos x$

Answer ____________

21. Given the equation of the curve $xy = 5 + y$, where y is the twice differentiable function of x, what is y''?

$(A)\ \dfrac{1-x+y}{(x-1)^2}$

$(B)\ \dfrac{2y}{(x-1)^2}$

$(C)\ -\dfrac{2y}{(x-1)^2}$

$(D)\ \dfrac{-y}{(x-1)}$

Answer_______________

22. A 25 feet ladder leans against a vertical wall. The foot of the ladder is pulled away from the wall horisontally at the rate of 6 ft / sec. What is the rate (in ft/sec) at which the top of the ladder is sliding down the wall when the top of the ladder is 15 feet above the ground?

$(A)\ -3$

$(B)\ -5$

$(C)\ -7$

$(D)\ -8$

Answer _______________

23. Which of the following are the absolute extrema of the function $f(x)=3x^2-9x+8$ on the interval $[-2,3]$?

(A) max $=8$, min $=\frac{5}{4}$

(B) max $=38$, min $=8$

(C) max $=38$, min $=\frac{5}{4}$

(D) max $=9$, min $=0$

Answer ______________

24. Shown below is the graph of the differentiable function $f(x)$. Which of the following statements about $f(x)$ must be true?

I. $f'(c)=0$ for some value c in the closed interval $[0,5]$

II. $f(c)=0$ for some value c in the open interval $(0,5)$

III. $f'(x)$ is an odd function

(A) I only (B) II only (C) I and II only (D) II and III only

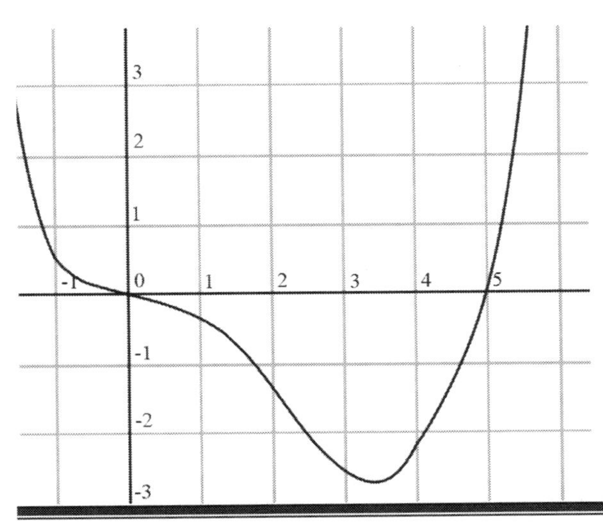

Answer ______________

25. How many local extrema does the function $f(x)=(x+1)^2(x-2)^3$ have on the open interval $(-9,5)$?

(A)One

(B)Two

(C)Three

(D)Four

Answer ______________________

26. A particle moves to the right along the x-axis until it reaches the origin and then moves along the x-axis to the left. The velocity of this particle is given by $v(t)=20-4t$ for $t\geq 0$. What is the position of the particle at any time t?

$(A)-2t^2+20t-50$

$(B)-2t^2+20t+50$

$(C)-4t^2+20t$

$(D)-4t^2+20t-50$

Answer ______________________

27. Let $f(t)=\frac{3}{t^2}$. For what value of t is $f'(t)$ equal to the average rate of change of f on a closed interval $[a,b]$?

$(A)\sqrt[3]{\frac{2a^2b^2}{a-b}}$ $(B)\sqrt[3]{\frac{2a^2b^2}{a+b}}$ $(C)-\sqrt[3]{\frac{2a^2b^2}{a+b}}$ $(D)\sqrt[3]{\frac{a^2b^2}{a+b}}$

Answer________________

28. Shown below is the slope field of which differential equation?

$(A)\ \frac{dy}{dx}=xy-y$

$(B)\ \frac{dy}{dx}=3y^2-4$

$(C)\ \frac{dy}{dx}=xy$

$(D)\ \frac{dy}{dx}=5x$

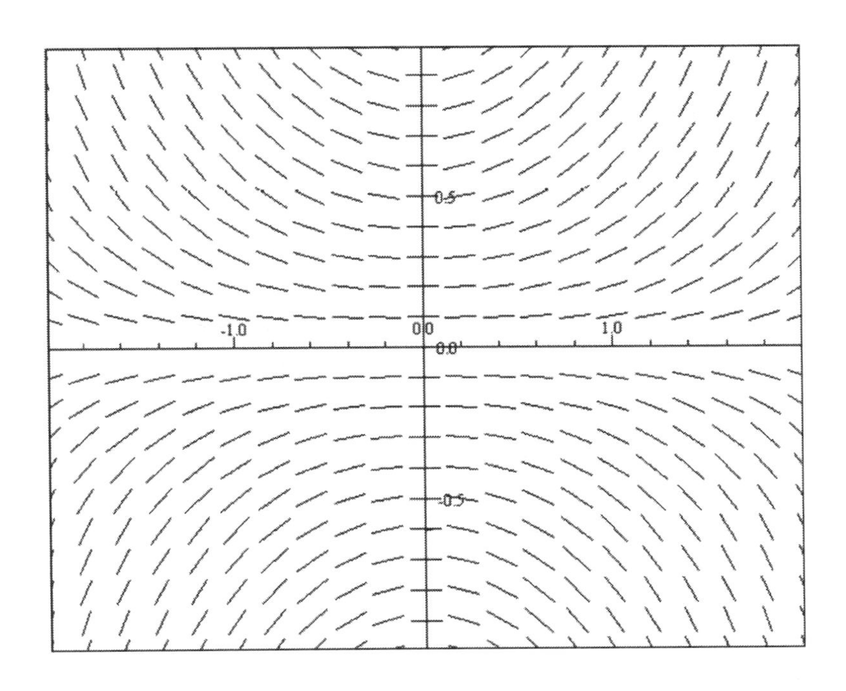

Answer________________

29. A solid has its base the region R bounded by the graph of $y=3\sqrt{x}$, the horizontal line $y=9$, and the y-axis. Which of the following is the integral expression that gives the volume of the solid if every cross section perpendicular to the y-axis is a rectangle whose height is 5 times the length of its base in region R?

$(A)\ 5\int_0^9 \left(3\sqrt{x}\right)^2 dx$ $(B)\ \int_0^9 \left(3\sqrt{x}\right)\left(15\sqrt{x}\right) dx$ $(C)\ \frac{5}{81}\int_0^9 y^4 dy$ $(D)\ \frac{5}{9}\int_0^9 y^2 dy$

Answer____________________

30. What is equation of the tangent line to the graph of $y=x-e^{-x}$ that is parallel to the line $6x-2y=15$?

$(A)\ y=3x+2\ln 2-2$ $(B)\ y=3x+2\ln 2+2$ $(C)\ y=3x-2\ln 2-2$ $(D)\ y=3x-2\ln 2+2$

Answer ____________________

Examination IV

Section I Part B

Directions: Solve each of the following problems, using the space provided. Choose the best answer. Do not spend too much time on any one problem. A graphing calculator is required for some questions on this part of the exam.

In this test:

1. The exact numerical value of the correct answer does not always appear among the choices given. Then select from among the choices the number that best approximates the exact numerical value.
2. Unless otherwise specified, the domain of a function is assumed to be the set of all real numbers x for which $f(x)$ is a real number.

31. What approximately is the area under the curve $f(x)=x^2+4x+6$ on the interval $[2,6]$ using the right-hand Riemann sum where P is the partition of $[2,6]$ determined by $\{2,4,5,6\}$?

(A) 193

(B) 160

(C) 195

(D) 143

Answer ____________________

32. The second derivative of function f is given by $f''(x)=\sin\left(e^{0.3x}\right)+\frac{x}{35}$. How many relative maximum points does $f'(x)$ have in the interval $0<x<10$?

(A) One

(B) Two

(C) Three

(D) More than three

Answer ____________________

33. Given $f''(x)=3+4\cos x$, $f'(0)=0$, and $f(0)=0$. The line, tangent to the graph of $f(x)$ and parallel to the segment connecting the endpoints of the interval $[0,5]$,touches $f(x)$ at $x=?$

$(A)0.596$ $(B)1.018$ $(C)1.381$ $(D)2.073$

Answer____________________

34. Given differentiable function $f(x)=2+\int_0^{2x}\sin(t^3)dt$. Which of the following is the smallest positive number c for which $f'(c)=0$?

(A) 0.327

(B) 0.463

(C) 0.656

(D) 0.732

Answer ____________________

35. Let R be the region bounded by the graphs: $y = 2x$, $y = kx$, $x = 0$, $x = 10$
If the area of R $= 50$ and k < 2, the value of k is :
(A)1.00
(B)1.25
(C)1.50
(D)1.75

Answer ____________________

36. Let $f(x)$ be a continuous and differentiable function on the interval $0 \le x \le 1$, and let $g(x) = f(2x)$. The table below gives values of $f'(x)$, the derivative of $f(x)$. What is the value of $g'(0.2)$?
(A) 2.32
(B) 3.01
(C) 4.64
(D) 6.02

x	0.1	0.2	0.3	0.4	0.5	0.6
f ' (x)	2.05	2.32	2.56	3.01	3.52	3.75

Answer ____________________

37. The girl runs along a straight horizontal road such that her position at any time t is given by $x(t)=2t^5+3t^2-4t^4+9$. For which value of t within [0, 1.3] is her speed the greatest?

(A) .31 (B) .45 (C) 1.13 (D) 1.27

Answer ____________

38. The velocity of a particle is given by $v(t)=t\sin t^2$. What is the distance traveled by the particle from time $t=0$ to time $t=\frac{\pi}{3}$?

(A) 0.225

(B) 0.235

(C) 0.272

(D) 0.342

Answer ____________

39. The graphs of $y = 2\sin x + 4$ and $y = -3\cos x + 5$ intersect each other on the interval $[1,5]$. The two regions bounded by these graphs and by the vertical lines $x = 1$ and $x = 5$ are revolved separately around the x-axis. What is the sum of the volumes of the two resulting solids of revolution?

(A) 91.200

(B) 359.131

(C) 574.622

(D) 851.890

Answer____________________

40. What is the volume of the solid generated when the region in the first quadrant bounded by the graphs of $y = x^3$ and $y = 4x$ is revolved around the line $x = 5$?

(A) 12.566

(B) 38.204

(C) 73.723

(D) 98.855

Answer ____________________

41. The acceleration of a particle moving along the line is given by $a(t)=t\cos(t^2)$. If at time $t=0\,\text{sec}$, its velocity is $2m/\text{sec}$ and position is $4m$, what is the position of the particle at time $t=7\,\text{sec}$?

(A) 14.303 m

(B) 4.303 m

(C) 17.697 m

(D) 18.303 m

Answer____________________

42. A wire 100 inches long is to be cut into two pieces. One of the pieces will be bent into a circle and the other into a square. What is the minimum sum of the areas of the circle and the square?

(A) 350.06 in^2 (B) 625 in^2 (C) 331.07 in^2 (D) 427.81 in^2

Answer ____________________

43. The graph of the function $f(x)$ is shown below. If $g(x)=\int_x^1 f(t)dt$, what is the limit of $g'(x)$ as x approaches 4?

(A) 6 (B) 3 (C) -2 (D) -3

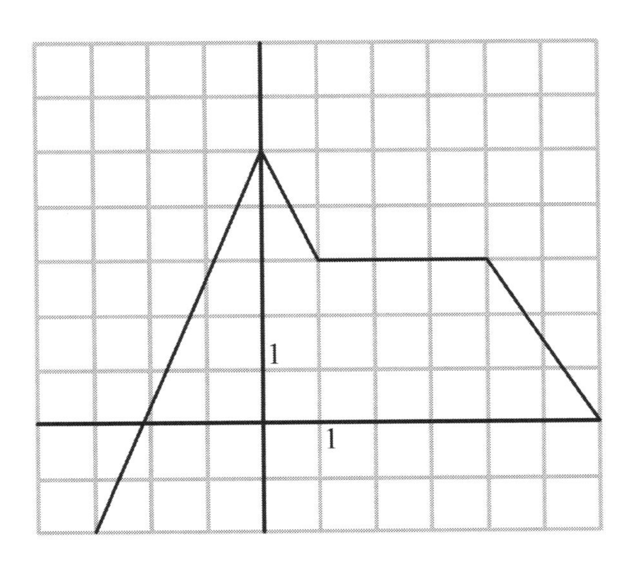

Answer ________________

44. At any time t $\geq$ 0, in days, the rate of growth of a mosquito population is given by $y' = ky$, where k is a constant of proportionality and y is the number of mosquitoes present. The initial population is 3,000 and the population quadruples during the first 7 days. By what factor will the population have increased in the first 14 days?

(A) 4

(B) 8

(C) 16

(D) 32

Answer ________________

45. Which of the following could be equal to $f(x)$, if it is given that $f'\left(\frac{\pi}{3}\right)=5$ and $f''(x)=8\cos x-\sin x$?

$(A)-8\cos x+\sin x-1.93x+7.25$

$(B)8\cos x-\sin x-1.93x+7.25$

$(C)8\cos x+\sin x-2.43x+7.25$

$(D)-8\cos x+\sin x-2.43x+7.25$

Answer________________

Examination V

Section I Part A

Directions: Solve each of the following problems, using the space provided. Choose the best answer. Do not spend too much time on any one problem. Calculators may NOT be used on this part of the exam.

In this test: Unless otherwise specified, the domain of a function is assumed to be the set of all real numbers x for which $f(x)$ is a real number.

1. $\lim\limits_{x\to 0} \dfrac{3-\sqrt{9+x}}{x} =$

(A) Does not exist

$(B)\, 6$

$(C) -\dfrac{1}{9}$

$(D) -\dfrac{1}{6}$

Answer ______________________

2. A function f satisfies the given conditions:

$\lim\limits_{x\to\infty} f(x)=0 \qquad \lim\limits_{x\to-\infty} f(x)=0 \qquad \lim\limits_{x\to 0^+} f(x)=\infty \qquad \lim\limits_{x\to 0^-} f(x)=\infty$

Which of the following is a possible graph for f, assuming that it does not cross a horizontal asymptote?

A)

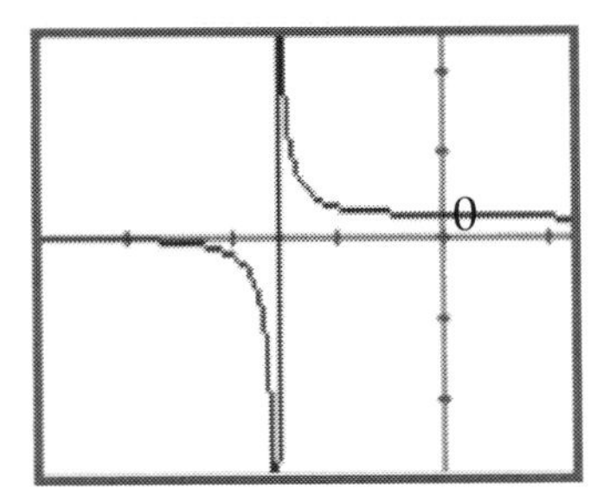

B)

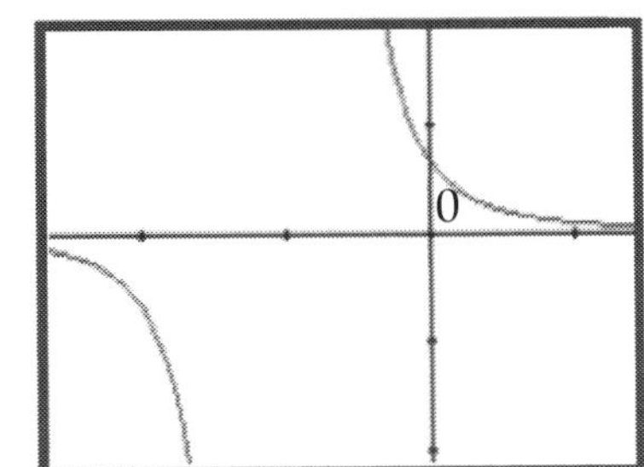

C)

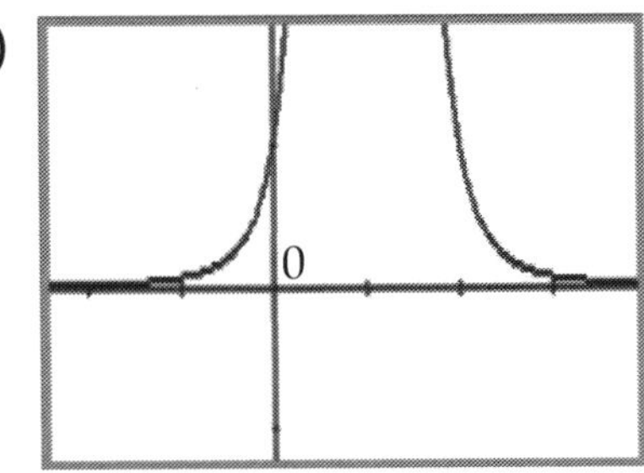

D)

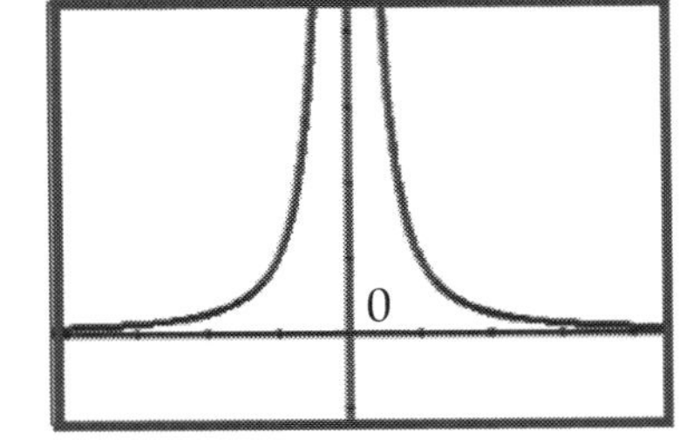

Answer ______________________

3. Let $y = f(x)$ be a particular solution to the differential equation $\frac{dy}{dx} = xy^3$ with $f(1) = 2$. What is the approximate value of $f(1.1)$ if the equation of the line tangent to the graph of $y = f(x)$ at $x = 1$ is being used to approximate $f(1.1)$?

(A) 3

(B) 2.2

(C) 2.6

(D) 2.8

Answer____________________

4. The graph of the function $f(x)$, consisting of two line segments, is shown in the figure below. Let g be the function given by $g(x) = x^2 + x - 1$, and let h be the function given by $h(x) = f(g(x))$. What is the value of $h'(1)$?

(A) 5 (B) $\frac{5}{3}$ (C) 3 (D) $\frac{3}{5}$

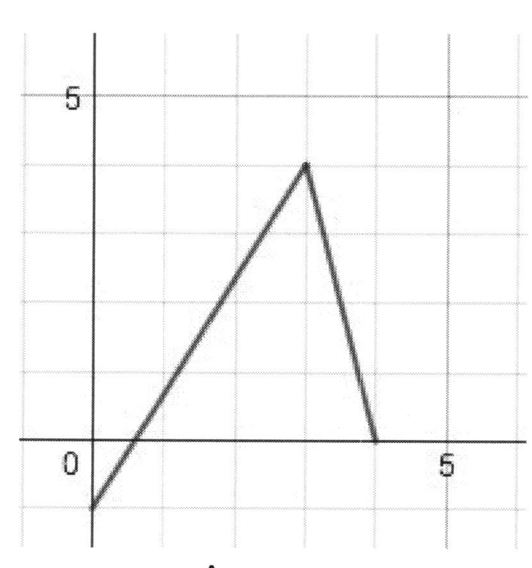

Graph of f

Answer____________________

x	0	4	6	8	16
$g(x)$	28	27	24	19	12
$g'(x)$	0	-1	-2	-3	-4

5. Let $g(x)$ be a differentiable function defined on the interval $0 \leq x \leq 16$. Some values of $g(x)$ and its derivative $g'(x)$ are given in the table above. Which of the following is the x - intercept of the line tangent to the graph of $g(x)$ and parallel to the segment connecting the endpoints of $g(x)$?

(A) $(23,0)$

(B) $(31,0)$

(C) $(-23,0)$

(D) $(-31,0)$

Answer____________________

6. What is y'' if $\sin y = y + 5x$?

(A) $\dfrac{5}{-1+\cos y}$ (B) $\dfrac{5\sin y}{(-1+\cos y)^2}$ (C) $-\dfrac{25\cos y}{(1+\cos y)^2}$ (D) $\dfrac{25\sin y}{(-1+\cos y)^3}$

Answer ____________________

7. Graph of $f''(x)$, the second derivative of function $f(x)$ is shown on the right. What must be true about $f(x)$?

I. $f(x)$ is concave downward for all $a < x < b$.

II. $f(x)$ must have a critical points at $x = a$ and $x = c$.

III. The slope of $f(x)$ increases when $x < a$.

(A) I only (B) III only (C) I and III only (D) I,II and III

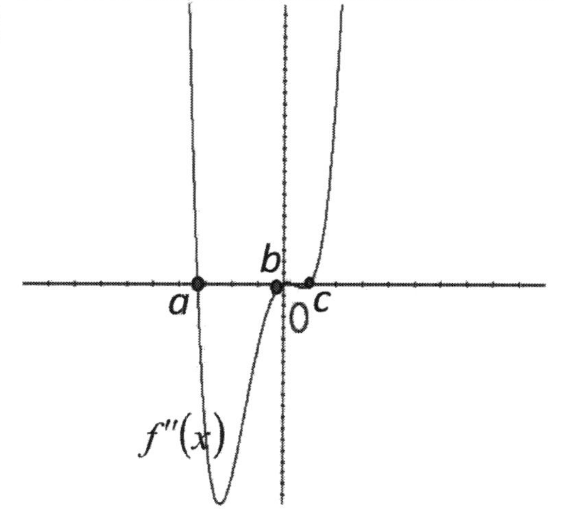

Answer____________________

8.At which value(s) of x does the function $f(x)=(x+1)^3(x-6)^2$ have a local minimum?

(A) 6 only

(B) 6 and -1 only

(C) $\frac{16}{5}$ and -1 only

(D) 6 and $\frac{16}{5}$ only

Answer ____________________

9. The curves $f(x)$, $f'(x)$, and $f''(x)$ are shown.

A function $f(x)$ is increasing and concave down.

What are the correct labels for graphs I, II, and III?

	I	II	III
(A)	$f'(x)$	$f(x)$	$f''(x)$
(B)	$f''(x)$	$f'(x)$	$f(x)$
(C)	$f(x)$	$f''(x)$	$f'(x)$
(D)	$f(x)$	$f'(x)$	$f''(x)$

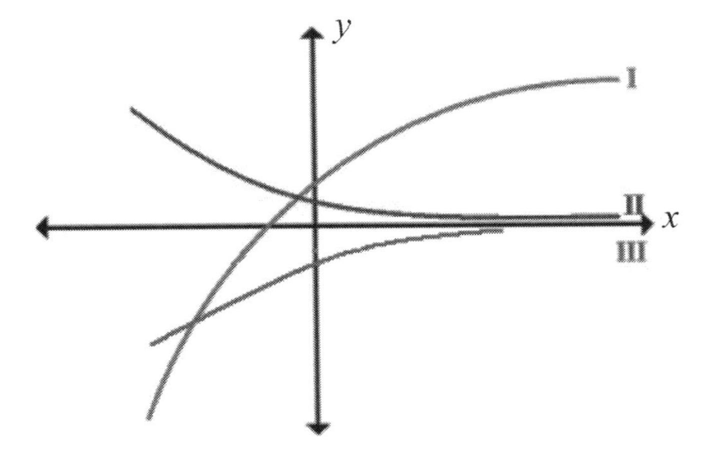

Answer ____________________

10. $\int \sec^2 \frac{x}{9} \tan \frac{x}{9}\, dx =$

$(A)\ \frac{1}{3}\tan^2\frac{x}{9}+C$

$(B)\ 9\tan^2\frac{x}{9}+C$

$(C)\ \frac{9}{2}\tan^2\frac{x}{9}+C$

$(D)\ \frac{9}{2}\tan\frac{x}{9}+C$

Answer ____________________

11. What is the area under the curve of $F(x) = f(x) - g(x)$ for $-3 \le x \le 3$ if $f(x) = 12 - g(x)$?

$(A)\ 72 - 2\int_{-3}^{3} g(x)dx$

$(B)\ 72 + \int_{-3}^{3} f(x)dx$

$(C)\ 72 + 2\int_{-3}^{3} g(x)dx$

$(D)\ \int_{-3}^{3} (6 - g(x))dx$

Answer ____________________

12. The graph of the velocity $v(t)$ of a particle is shown below. What is the total distance traveled by the particle from time $t = 0$ to $t = 5$?

$(A) 2\pi - 1.5$ $(B) 4\pi + 1.5$ $(C) 2\pi + 1.5$ $(D) 2\pi + \sqrt{10}$

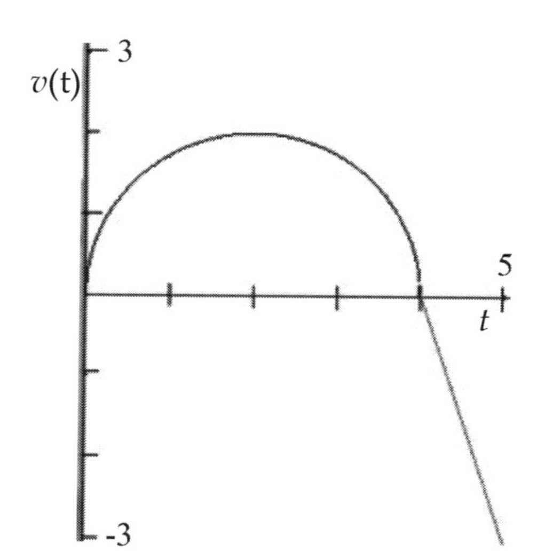

Answer ____________________

13. The region bounded by the graphs of $y=-\sin x$, $y=\sin x$, $x=0$, and $x=\pi$ is revolved around $y=2$. What is the volume of the resulting solid?

$(A)\,8\pi$ $(B)\,16\pi$ $(C)\,4\pi$ $(D)\,32\pi$

Answer ______________________

14. The position $s(t)$ of a particle on the x-axis at time t, $t\geq 0$, is $\cos t$. The average velocity of the particle for $0\leq t\leq \pi$ is

$(A)\,-\frac{1}{\pi}$ $(B)\,\frac{2}{\pi}$ $(C)\,-\frac{2}{\pi}$ $(D)\,\frac{1}{\pi}$

Answer ______________________

15. Which of the following is equal to the area of the region bounded by $y=(x-1)^2$ and $y=x+1$?

$(A)\ \frac{9}{2}$ $(B)\ 3$ $(C)\ 18$ $(D)\ \frac{11}{2}$

Answer ________________

16. A particle moves on the x-axis so that its position at any time $t \geq 0$ is given by $x(t)=4te^{-2t}$. What is the velocity of the particle at time $t=3$?

$(A)\ 28e^{-6}$ $(B)\ -20e^{-6}$ $(C)\ 12e^{-6}$ $(D)\ -12e^{-6}$

Answer ________________

17. The graphs of two differentiable functions $p(x)$ and $q(x)$ are shown below. Given $h(x) = p(x)q(x)$, which of the following statements about $h'(-1)$ is true?

(A) $h'(-1) < 0$ (B) $h'(-1) > 0$ (C) $h'(-1) = 0$

(D) There is not enough information given to conclude anything about $h'(-1)$

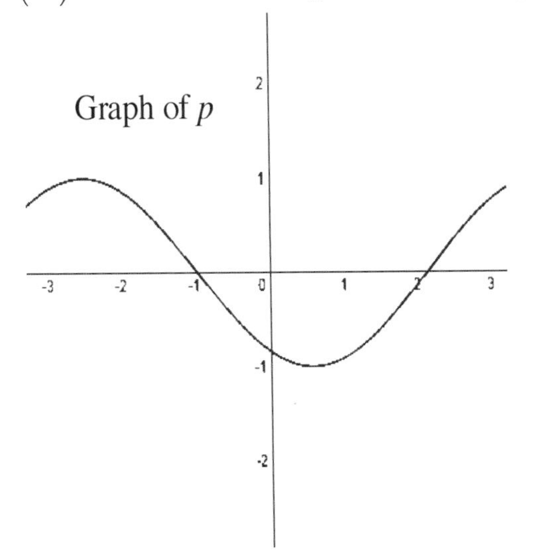

Graph of q

Answer ____________________

18. A particle moves along the x-axis so that its velocity at any time $t \geq 0$ is given by $v(t) = \pi \sin(\pi t)$. What is the total distance the particle travels from $t = 0$ to $t = 2$?

(A) 0 (B) 2 (C) 2π (D) 4

Answer ____________________

19. What is the the volume of the solid generated when the region in the first quadrant bounded by the graph of the function $y=3x-x^2$ is revolved around the $x-$axis?

$(A)\frac{9\pi}{2}$

$(B)\frac{81\pi}{10}$

$(C)\frac{32\pi}{5}$

$(D)\frac{81}{10}$

Answer____________________

x	-1	2	4	6	7
f (x)	3	5	-2	8	-6
f′(x)	2	6	9	7	5

20. Let $g(x)$ be the inverse of the function $f(x)$. Given the following values on the table above, at which value $x=a$ will $g'(a)=\frac{1}{6}$?

$(A)-1$

$(B)\ 2$

$(C)\ 3$

$(D)\ 5$

Answer ____________________

21. Which of the following is the solution to the differential equation $\frac{dy}{dx}=\frac{x-5}{y+1}$; $y\neq -1$, subject to initial conditions $y(0)=3$?

$(A)\, y = x-6$

$(B)\, y=\sqrt{(x-5)^2-9}-1$

$(C)\, y=\sqrt{(x-5)^2-9}$

$(D)\, y=\sqrt{(x-5)^2-16}$

Answer ____________________

22. If $\int_{60}^{150} v(t)dt = 127$, and $\int_{80}^{150} v(t)dt = -78$, what is $\int_{60}^{80} v(t)dt$?

(A) 49 (B) 205 (C) -205 (D) -49

Answer ____________________

23. A projectile is fired from 16 ft above the ground with an initial velocity of 10 ft/s. Its downward acceleration is represented by $a(t) = -12t^3$. Which of the following represents the height of the projectile at time t, if time $t = 0$ when it is launched?

$(A) -\frac{4}{5}t^5 + 10t + 16$

$(B) -\frac{3}{5}t^5 + 10t + 16$

$(C) -3t^4 + 10$

$(D) -\frac{3}{5}t^5 - 10t + 16$

Answer________________

24. The tangent line to the curve $y = e^{5-2x}$ at the point $(2, e)$ intersects the x and y axis. What is the area of the triangle formed by the tangent line, the x-axis and the y-axis?

$(A)\frac{25}{4}e$ $(B)\frac{5}{4}e$ $(C)\frac{25}{2}e$ $(D)5e$

Answer________________

25. Shown below is the slope field for which differential equation?

$(A)\ \dfrac{dy}{dx} = 1 + y^2$

$(B)\ \dfrac{dy}{dx} = x - y$

$(C)\ \dfrac{dy}{dx} = 1 + x^2$

$(D)\ \dfrac{dy}{dx} = 1 - y^2 + x^2$

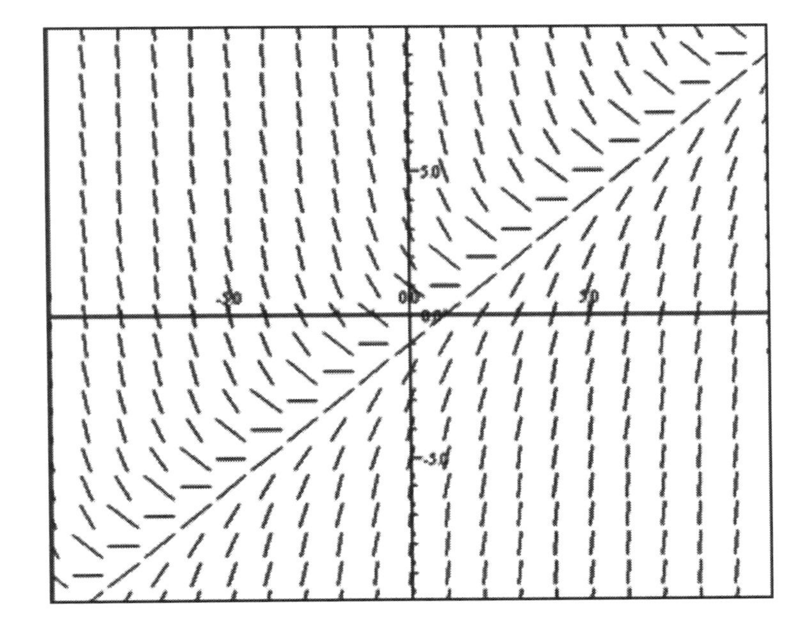

Answer ____________________

26. If $y = \ln\sqrt[4]{\dfrac{3x^2 - 3}{x^3 + 5}}$ what is $\dfrac{dy}{dx}$?

$(A)\ \sqrt[4]{\dfrac{x^3 + 5}{3x^2 - 3}}$

$(B)\ \dfrac{3x^2 - 3}{4(x^3 + 5)}$

$(C)\ \dfrac{x(3x + 10 - x^3)}{4(x^3 + 5)(x^2 - 1)}$

$(D)\ \dfrac{6x(3x^2 - 3)}{4(x^3 + 5)}$

Answer ____________________

27. The base of a solid is the region in the first quadrant bounded by the line $x + y = 4$ and the coordinate axes. What is the volume of the solid if every cross section perpendicular to the x-axis is a semicircle?

$(A)\frac{4\pi}{3}$ $(B)\frac{8\pi}{3}$ $(C)\frac{32\pi}{3}$ $(D)\frac{64\pi}{3}$

Answer________________

28. $\int_2^3 \frac{x^2}{2x^3 - 7}\,dx = ?$

$(A)\ln 38$ $(B)\frac{1}{6}\ln\frac{47}{9}$ $(C)\frac{1}{6}\ln 38$ $(D)\ln\frac{47}{9}$

Answer________________

29. What is the volume of the solid generated when the region bounded by the graphs of $x=\sqrt{y}$ and $x=\frac{1}{3}y$ is rotated around the y-axis?

$(A)\ \frac{9\pi}{2}$ $(B)\ 9\pi$ $(C)\ \frac{27\pi}{2}$ $(D)\ 18\pi$

Answer ______________

30. Let $f(x)=\begin{cases} 2x-e^{x-5} & 0\le x\le 15 \\ 2x-e^{25-x} & 15<x\le 20 \end{cases}$

Which of the following is true?

I. $f(x)$ is continuous for all values of x in the interval $[0,20]$.

II. $f'(x)$ is continuous for all values of x in the interval $[0,20]$.

III. $f(x)$ is concave down for all values of x in the interval $[0,15)\cup(15,20]$

(A) I only (B) I and III only (C) II and III only (D) I, II, and III

Answer ______________

Examination V

Section I Part B

Directions: Solve each of the following problems, using the space provided. Choose the best answer. Do not spend too much time on any one problem. A graphing calculator is required for some questions on this part of the exam.

In this test:

1. The exact numerical value of the correct answer does not always appear among the choices given. Then select from among the choices the number that best approximates the exact numerical value.
2. Unless otherwise specified, the domain of a function is assumed to be the set of all real numbers x for which $f(x)$ is a real number.

31. Water leaks from a pipe into a bucket at a rate of $1.7+\frac{4.5}{x}$, x being the time in minutes that the pipe has been leaking. If there is 16.52 mL of water in the bucket after 2 minutes, how much water is in the bucket after 5 minutes?

(A) 25.74 mL (B) 24.67 mL (C) 20.12 mL (D) 16.70 mL

Answer________________

32. The temperature of a cup of tea is modeled by function H for $0 \le t \le 15$. Values of $H(t)$ for selected values of t are shown in the table below. Using a trapezoidal sum with the 5 subintervals indicated by the table below, what is the approximate average temperature of the tea over 15 minutes.

(A) 58.333 °C

(B) 57.513 °C

(C) 55.833 °C

(D) 53.333 °C

t(min)	0	3	4	8	10	15
H(t) (degrees C)	70	65	60	55	50	45

Answer ________________

33. At $3:00$ PM car A is 50 miles south of car B and is driving north at a rate of 25mph. If car B is driving west at a rate of 15 mph, at what time is the distance between the cars minimal?

$(A)4:12$ PM (B) $4:28$ PM $(C)4:45$ PM $(D)3:44$ PM

Answer ____________________

34. What is the average value of the function $f(x)=150\sin\left(\frac{2\pi}{365}(x-80)\right)+670$ on the interval $[1,181]$?

(A) 688.22

(B) 714.87

(C) 345.44

(D) 1439.62

Answer ____________________

35. Let R be the region enclosed by $y = 3\ln x$, $x = 5$ and $x-$axis. What is the volume of the solid generated by revolving the region R about the x-axis?

(A) 193.879

(B) 137.330

(C) 149.532

(D) 187.913

Answer____________________

36. Cream puffs are being filled with custard at the rate $r(t) = 80t^{\frac{1}{4}}$ cream puffs per minute, with $t \geq 0$ measured in minutes. If there were 700 filled cream puffs to begin with, how many cream puffs are filled after 625 minutes?

(A)100700 (B)200000 (C)200700 (D)300000

Answer ____________________

37. Let $f(t) = \arcsin(t)$ for $0 \le t \le 1$. For what value of t is the instantaneous rate of change of $f(t)$ equal to the average rate of change of $f(t)$ on the closed interval $[0.5, 1]$?

(A) 0.4726 (B) 0.8787 (C) 0.6868 (D) 0.9435

Answer ______________________

38. What is the volume generated when the region bounded by $y = \tan x, y = \cos x$, and the y-axis is revolved about the x-axis?

(A) 0.792 (B) 1.186 (C) 1.246 (D) 1.433

Answer ______________________

39. Which of the following is the best approximation of the area of the region between $y = \ln x$ and $y = \log x$ from their point of intersection to $x = 4$?

(A) 1.326 (B) 1.440 (C) 1.513 (D) 1.682

Answer________________

40. Consider the area of the region bounded by the graph of $y = \left(x - \frac{1}{2}\right)^2 + \frac{1}{2}$ and the x-axis in the interval $[0,2]$. Arrange the following in increasing order:

(I) Area calculated using Right Riemann Sums with four subintervals

(II) Area calculated using Left Riemann Sums with four subintervals

(III) Area calculated by evaluating the integral.

(A) I, III, II

(B) I, II, III

(C) II, I, III

(D) II, III, I

Answer________________

41. The derivative of function $f(x)$ is given by $f'(x)=e^{x}\left(-x^{5}+4x\right)-7$. At what value of x does $f(x)$ have an absolute minimum on the interval $\left[0,3\right]$?

$(A)3$ $(B)0$ $(C)0.858$ $(D)1.240$

Answer ______________________

42. The base of a container is defined by the region bounded by the graph of $y=-x^{2}+6x+9$ in the first quadrant. The cross-sections of the container are semicircles perpendicular to the x-axis. Which of the following represents the volume of the container?

$(A)2\pi\int_{0}^{7.24264}\left(\frac{1}{2}\left(-x^{2}+6x+9\right)\right)^{2}dx$

$(B)\frac{\pi}{2}\int_{0}^{7.24264}\left(\frac{1}{2}\left(-x^{2}+6x+9\right)\right)^{2}dx$

$(C)\pi\int_{0}^{7.24264}\left(\frac{1}{2}\left(-x^{2}+6x+9\right)\right)^{2}dx$

$(D)\frac{1}{2}\int_{-1.24264}^{7.24264}\left(-x^{2}+6x+9\right)^{2}dx$

Answer ______________________

43. If the function g is defined by $g(x)=\int_0^{x^2}\cos t\,dt$ on the closed interval $-\frac{1}{2}\le x\le\frac{5}{2}$, then g has a local maximum at $x=$

(A) 0

(B) 0.808

(C) 1.253

(D) 2.171

Answer____________________

44. What is the area in the first quadrant of the region R, bounded by the y-axis, horizontal line $y=2$ and $y=\sqrt{x^3+1}$?

(A) 0.7587 (B) 1.0262 (C) 1.4423 (D) 3.2450

Answer ____________________

45. The amount y of a radioactive substance decays according to the equation $\frac{dy}{dt} = ky$ where k is a constant and time, t, is measured in days. If half of the amount present will decay in 13 days, what is the value of k?

$(A) -0.053$

$(B) -0.015$

$(C) -0.072$

$(D) -0.026$

Answer________________________

Formulas and Theorems

Limits

$\lim\limits_{x\to c}\left[f(x)+g(x)\right]=\lim\limits_{x\to c}f(x)+\lim\limits_{x\to c}g(x)$	$\lim\limits_{x\to c}\left[af(x)\right]=a\lim\limits_{x\to c}f(x)$
$\lim\limits_{x\to c}\left[f(x)\cdot g(x)\right]=\lim\limits_{x\to c}f(x)\cdot\lim\limits_{x\to c}g(x)$	$\lim\limits_{x\to c}\left[f(x)^{\frac{a}{b}}\right]=\left[\lim\limits_{x\to c}f(x)\right]^{\frac{a}{b}}$
$\lim\limits_{x\to c}\left[\dfrac{f(x)}{g(x)}\right]=\dfrac{\lim\limits_{x\to c}f(x)}{\lim\limits_{x\to c}g(x)}$	$\lim\limits_{n\to 0}(1+n)^{\frac{1}{n}}=\lim\limits_{n\to\infty}\left(1+\dfrac{1}{n}\right)^{n}=e$
$\lim\limits_{x\to 0}\dfrac{\sin x}{x}=\lim\limits_{x\to 0}\dfrac{x}{\sin x}=1$	$\lim\limits_{x\to 0}\dfrac{1-\cos x}{x}=0$

Differentiation formulas

Power Rule: $\left(x^{n}\right)'=nx^{n-1}$	Product Rule: $\left(f(x)\cdot g(x)\right)'=f'(x)\cdot g(x)+g'(x)\cdot f(x)$
Reciprocal Rule: $\left(\dfrac{1}{g(x)}\right)'=-\dfrac{g'(x)}{g^{2}(x)}$	Quotient Rule: $\left(\dfrac{f(x)}{g(x)}\right)'=\dfrac{f'(x)\cdot g(x)-g'(x)\cdot f(x)}{g^{2}(x)}$
$(\sin x)'=\cos x$	Chain Rule: $\left(f(g(x))\right)'=f'(g(x))\cdot g'(x)$
$(\cos x)'=-\sin x$	$(a)'=0$
$(\tan x)'=\sec^{2}x$	$(x)'=1$
$(\cot x)'=-\csc^{2}x$	$\left(e^{x}\right)'=e^{x}$
$(\sec x)'=\sec x\tan x$	$(\ln x)'=\dfrac{1}{x}$
$(\csc x)'=-\csc x\cot x$	$\left(a^{x}\right)'=a^{x}\ln a,\ a>0,\ a\neq 1$
$\left(\sin^{-1}x\right)'=\dfrac{1}{\sqrt{1-x^{2}}}$	$\left(\log_{a}x\right)'=\dfrac{1}{x\ln a}$
$\left(\tan^{-1}x\right)'=\dfrac{1}{1+x^{2}}$	$\left(\sqrt{x}\right)'=\dfrac{1}{2\sqrt{x}}$
$\left(\sec^{-1}x\right)'=\dfrac{1}{x\sqrt{x^{2}-1}}$	$\left(\dfrac{1}{x}\right)'=-\dfrac{1}{x^{2}}$

Formulas and Theorems

Integration Formulas

$\int \sin x\,dx = -\cos x + C$	$\int x^n\,dx = \frac{x^{n+1}}{n+1} + C,\ n \neq -1$
$\int \cos x\,dx = \sin x + C$	$\int a\,dx = ax + C$
$\int \tan x\,dx = -\ln\lvert\cos x\rvert + C$ or $\ln\lvert\sec x\rvert + C$	$\int e^x\,dx = e^x + C$
$\int \cot x\,dx = \ln\lvert\sin x\rvert + C$ or $-\ln\lvert\csc x\rvert + C$	$\int a^x\,dx = \frac{a^x}{\ln a} + C$
$\int \sec x\,dx = \ln\lvert\sec x + \tan x\rvert + C$	$\int \frac{1}{x}\,dx = \ln\lvert x\rvert + C$
$\int \csc x\,dx = \ln\lvert\csc x - \cot x\rvert + C$	$\int \frac{1}{\sqrt{a^2 - x^2}}\,dx = \sin^{-1}\frac{x}{a} + C$
$\int \sec x \tan x\,dx = \sec x + C$	$\int \frac{1}{a^2 + x^2}\,dx = \frac{1}{a}\tan^{-1}\frac{x}{a} + C$
$\int \csc x \cot x\,dx = -\csc x + C$	$\int \frac{1}{x\sqrt{x^2 - a^2}}\,dx = \frac{1}{a}\sec^{-1}\frac{x}{a} + C$
$\int \sec^2 x\,dx = \tan x + C$	$\int \csc^2 x\,dx = -\cot x + C$

Continuity	A function $f(x)$ is continuous at $x = a$ if all of the following are true: I. $f(a)$ exists II. $\lim_{x \to a} f(x)$ exists III. $\lim_{x \to a} f(x) = f(a)$
Intermediate Value Theorem	If function $f(x)$ is continuous on a closed interval $[a,b]$ and if w is any number between $f(a)$ and $f(b)$, then there is at least one number c in $[a,b]$ such that $f(c) = w$
Limit Theorem	$\lim_{x \to a} f(x) = L$ if and only if $\lim_{x \to a^+} f(x) = L = \lim_{x \to a^-} f(x)$

Formulas and Theorems

Vertical Asymptote	A line $x=a$ is a vertical asymptote of the graph of a function $y=f(x)$ if $\lim_{x\to a^+} f(x)=\pm\infty$ or $\lim_{x\to a^-} f(x)=\pm\infty$
Horizontal Asymptote	A line $y=L$ is a horizontal asymptote of the graph of a function $f(x)$ if $\lim_{x\to\infty} f(x)=L$ or $\lim_{x\to-\infty} f(x)=L$
Average Rate of Change	Average rate of change of $y=f(x)$ on $[a,a+h]$ is $\dfrac{f(a+h)-f(a)}{h}$
Instantaneous Rate of Change	Instantaneous rate of change of $y=f(x)$ on $[a,a+h]$is: $\lim_{h\to 0}(\text{Average rate of change})=\lim_{h\to 0}\dfrac{f(a+h)-f(a)}{h}$
Sandwich Theorem	If $f(x)\le g(x)\le h(x)$ for all $x\ne c$ in some interval about c,and $\lim_{x\to c} f(x)=\lim_{x\to c} h(x)=L$, then $\lim_{x\to c} g(x)=L$
Definition of Derivative	$f'(x)=\lim_{h\to 0}\dfrac{f(x+h)-f(x)}{h}$ or $f'(x)=\lim_{x\to c}\dfrac{f(x)-f(c)}{x-c}$
Theorem	If a function $f(x)$ is differentiable at $x=a$, then it is continuous at $x=a$
Differentiability on a closed interval	A function f is differentiable on a closed interval $[a,b]$ if f is differentiable on the open interval $(a,b)$and if both Right-Hand Derivative at a and Left-Hand Derivative at b exist
Rolle's Theorem	Suppose that $f(x)$ is continuous on the closed interval $[a,b]$ and differentiable on the open interval (a,b). If $f(a)=f(b)$, then there is at least one number c between a and b such that $f'(c)=0$
Mean Value Theorem	If $f(x)$ is continuous on the closed interval $[a,b]$ and differentiable on the open interval (a,b), then there is at least one number c between a and b such that $f'(c)=\dfrac{f(b)-f(a)}{b-a}$
Derivative of the Inverse function	$g'(c)=\dfrac{1}{f'(g(c))}$
Linear approximation of f(x) near x = x_0	$y=f(x_0)+f'(x_0)(x-x_0)$

Formulas and Theorems

Extrema and Concavity

Critical Numbers (Candidates for Extrema)	A number c in the domain of a function f is a critical number of f if either $f'(c)=0$ or $f'(c)$ does not exist
Increasing/Decreasing Functions	If $f(x)$ is differentiable on (a,b) and continuous on $[a,b]$: $f'(x)>0$ on $(a,b) \Rightarrow f(x)$ is increasing on $[a,b]$ $f'(x)<0$ on $(a,b) \Rightarrow f(x)$ is deccreasing on $[a,b]$
Local Minimum	$f(c)$ is the local minimum value of f if $f(c) \le f(x)$ for every x in an interval I around c
Local Maximum	$f(c)$ is the local maximum value of f if $f(c) \ge f(x)$ for every x in an interval I around c
Absolute Minimum	$f(c)$ is the absolute minimum value of f if $f(c) \le f(x)$ for every x in the domain of f
Absolute Maximum	$f(c)$ is the absolute maximum value of f if $f(c) \ge f(x)$ for every x in the domain of f
Finding the Absolute Extrema on [a,b]	1. Find Critical numbers. 2. Calculate the function values at the endpoints of $[a,b]$ and at the critical numbers. 3. The largest of these function values is the absolute maximum The smallest function value is the absolute minimum.
Finding the Local Extrema (First Derivative Test)	Find critical numbers $x=c$ (where $f'(c)=0$ or DNE) Local minimum occurs at $x=c$, where $f'(x)$ changes from negative to positive. Local maximum occurs at $x=c$, where $f'(x)$ changes from positive to negative .
Concavity	If $f''(x)$ exists on (a,b), then: $f''(x)>0 \Rightarrow f(x)$ concave upward in (a,b) $f''(x)<0 \Rightarrow f(x)$ concave downward in (a,b)
Points of Inflection	Points of Inflection of $f(x)$ are the points on the Domain of $f(x)$ where $f''(x)=0$ or DNE and $f''(x)$ changes it's sign passing through them.
Extreme Value Theorem	If f is continuous on a closed interval $[a,b]$, then f has both an absolute minimum and an absolute maximum value on $[a,b]$

Formulas and Theorems

Integrals

Definite Integral. Property 1	$\int_a^b c\cdot f(x)dx = c\int_a^b f(x)dx$
Definite Integral. Property 2	$\int_a^b f(x)dx = -\int_b^a f(x)dx$
Definite Integral. Property 3	$\int_a^a f(x)dx = 0$
Definite Integral. Property 4	If f is integrable on a closed interval and if a, b and c are any three numbers in the interval, then $\int_a^b f(x)dx = \int_a^c f(x)dx + \int_c^b f(x)dx$
If $f(x)\geq 0$ on $[a,b]$:	If $f(x)\geq 0$ on $[a,b]$ then $\int_a^b f(x)dx \geq 0$
If $g(x)\geq f(x)$ on$[a,b]$:	If $g(x)\geq f(x)$ on$[a,b]$ then $\int_a^b f(x)dx \geq 0$
If f(x) is an even function:	If f is an Even Function, then $\int_{-a}^a f(x)dx = 2\int_0^a f(x)dx$
If f(x) is an odd function:	If f is an Odd Function, then $\int_{-a}^a f(x)dx = 0$
Riemann sum	Let $f(x)$ be defined on a closed interval $[a,b]$and P is any decomposition of $[a,b]$into subintervals of the form $[x_{k-1},x_k]$. Riemann sum of $f(x)$ for P is $R_P = \sum_{k=1}^{n} f(w_k)\Delta x_k$ where $w_k \in [x_{k-1},x_k]$.
Integral as limit of Riemann sums	The definite integral of f from a to b, $\int_a^b f(x)dx = \lim_{\|p\|\to 0}\sum_k f(w_k)\Delta x_k$ provided the limit exists where $\|p\|$ is the norm of the partition(largest Δx_k)
Integral as area under the curve	If f is integrable and $f(x)\geq 0$ for every x in $[a,b]$(where $a<b$),then the area A of the region under the graph of f from a to b is $A=\int_a^b f(x)dx$
Fundamental Theorem of Calculus	$\frac{d}{dx}\int_a^x f(t)dt = f(x)$ and $\int_a^b f'(x)dx = f(b)-f(a)$
Velocity	$v(t)=x'(t)$ $v(t)=\int a(t)dt$

Formulas and Theorems

Acceleration	$a(t)=v'(t)=x''(t)$
Speed	$\text{Speed}=\lvert v(t)\rvert=\sqrt{\left(\frac{dx}{dt}\right)^2+\left(\frac{dy}{dt}\right)^2}$
Position	$x(t)=\int v(t)dt$
Average Velocity	$v_{av}=\frac{s(t_2)-s(t_1)}{t_2-t_1}$
Instantaneous velocity	$v_{inst}=\lim_{t_2\to t_1}\frac{s(t_2)-s(t_1)}{t_2-t_1}$
Total Distance	$\text{Total Distance}=\int_a^b \lvert v(t)\rvert dt$
Net Change	$\text{Net Change}=\int_a^b v(t)dt$
Area between the curves	$A=\int_a^b (f(x)-g(x))dx$, where $f(x)>g(x)$ on $[a,b]$
Volume by Disks	$V=\pi\int_a^b R^2\,dx$ (if region is rotating around horizontal line) or: $V=\pi\int_c^d R^2\,dy$ (if region is rotating around vertical line)
Volume by Washers	$V=\pi\int_a^b (R^2-r^2)dx$ (if region is rotating around horizontal line) or: $V=\pi\int_c^d (R^2-r^2)dy$ (if region is rotating around vertical line)
Volume by Cross Sections	$V=\int_a^b A(x)dx$ (if the cross sections are $\perp$ to x-axis) or: $V=\int_c^d A(y)dy$ (if the cross sections are $\perp$ to y-axis)
Average value of f	Average Value of $f(x)$ on $[a,b]$ is: $f_{av}=\frac{1}{b-a}\int_a^b f(x)dx$
Power-reducing trig formulas	$\sin^2 x=\frac{1-\cos 2x}{2}$ and $\cos^2 x=\frac{1+\cos 2x}{2}$
L'Hopital's Rule	If $\lim_{x\to c}\frac{f(x)}{g(x)}$ is of the form $\frac{0}{0}$ or $\frac{\infty}{\infty}$, then $\lim_{x\to c}\frac{f(x)}{g(x)}=\lim_{x\to c}\frac{f'(x)}{g'(x)}$, provided either $\lim_{x\to c}\frac{f'(x)}{g'(x)}$ exists or $\lim_{x\to c}\frac{f'(x)}{g'(x)}=\infty$

Tips for the AP Test

- Functions could have various notations such as: $Si(x)$, $erf(x)$
- Don't cross out your work unless you know you can do better. But if you make a mistake, cross it out. A crossed out solution is not graded.
- If you are afraid your result is wrong in Part A, use it anyway to finish the problem.
- Don't write $f(x)=2(1.5)+3$ when you mean $f(1.5)=2(1.5)+3$.
- Name the function you are referring to. Not "its slope is…" but "the slope of g is…" when more than one function is being used.
- If you are being asked to set up an integral and you are not sure about integrand, start with at least the limits of integration and constant in front of it and make a guess at the integrand.
- Know the difference between local (relative) and global (absolute) extrema.
- Know the difference between the **extreme value** and the **location** of the extreme value. If they ask you to "find the *minimum*", they want the *y-value*. If they ask, "At which *x* does this function have a minimum?" they want the *x-value*. If they ask, "At which *point* does this function have a minimum?" they want *both coordinates (x,y)*.
- Know the difference between a **point in time** and an **interval of time**. "During, before", "within", "after", and "until" are intervals. "At" or "when" indicate a point in time!
- If you use a rule, show how it applies to the given problem. Generic theorems or procedures get no credit. Don't write "By Mean Value Theorem" etc.; instead explain how the conditions and rules apply to your problem.
- Volume problems that are given **will not** require use of the **Shell method**. Shell method can be used, but problem could also be solved using Disc, Washers, or Cross Section method.
- If in Washer method you write $(f(x)-g(x))^2$ vs. $(f(x))^2-(g(x))^2$ you might **LOSE ALL POINTS & NOT BE GIVEN ANY PARTIAL CREDIT.**
- In rotation problems, when the solid is rotating around a Horizontal line $y = 2$, or Vertical line $x = 1$, be careful about rotating the figure around the **correct** line.

- Answers have to be reasonable. If you make a mistake in finding the area under the curve, and the answer is negative, it is UNREASONABLE. You can lose points for such errors and NOT BE GIVEN PARTIAL CREDIT.

- If you have "=" sign make sure that left is equal to the right. Don't put something like $2+5=7+1=8$. Instead put $2+5=7;\ 7+1=8$.

- Point-slope form of an equation, $y-y_0=m(x-x_0)$, is perfectly acceptable. There is no need to simplify it into $y=mx+b$ form. The form $\dfrac{y-y_0}{x-x_0}=m$ is not acceptable because of the restrictions for the denominator.

- Answers like $2+\sin\dfrac{\pi}{3},\ 2+\tan^{-1}1,\ \dfrac{(.1)^2x^2}{2!}+\dfrac{(.1)^3x^3}{3!}$ are acceptable. You don't have to simplify it anymore.

- When you use trapezoidal approximation, your answer is a complicated sum. Leave it as a sum, like 2.71 + 3.06 + 5.06. You don't have to do the arithmetic and add the terms up.

- THEY DON'T ACCEPT diagrams for justification of local maximum, local minimum, and concavity. Instead you need to write the explanation similar to: **Local minimum occurs at x = 3, because that's where derivative of *f*** (or *g* or *h* etc.) **changes from negative to positive.**

- For every part of the free response, give all the answers with at least 3 digits after the decimal point, rounded or truncated. More is okay. Make sure the numbers you are using for earlier steps of the solution contain MORE THAN 3 digits after the decimal point. Don't round too soon, or else your answer may be off.

- When graphing the slope field (direction fields), extend it all the way through the given window.

Answer Key

	Exam 1	Exam 2	Exam 3	Exam 4	Exam 5
1)	C	C	C	B	D
2)	A	C	B	D	D
3)	B	C	D	B	D
4)	C	B	A	C	A
5)	C	D	A	A	B
6)	B	D	A	C	D
7)	A	B	B	C	C
8)	C	D	B	C	A
9)	C	B	C	C	D
10)	B	A	C	B	C
11)	A	D	C	A	A
12)	B	D	B	D	C
13)	C	B	D	C	B
14)	A	A	B	C	C
15)	C	D	A	A	A
16)	B	C	A	A	B
17)	C	A	A	B	A
18)	A	B	C	A	D
19)	C	C	D	C	B
20)	D	C	A	D	D
21)	D	B	C	B	B
22)	B	C	B	D	B
23)	D	C	B	C	B
24)	B	D	D	A	A
25)	C	D	B	B	B
26)	C	B	D	A	C
27)	B	C	A	B	B
28)	B	B	A	C	B
29)	B	B	B	C	C
30)	B	D	B	A	B
31)	B	B	A	A	A
32)	A	A	C	C	C
33)	C	C	A	C	B
34)	A	C	A	D	A
35)	D	A	B	A	B
36)	C	D	A	D	C
37)	C	C	B	B	B
38)	A	C	B	C	D
39)	C	C	C	B	B
40)	D	D	A	D	D
41)	A	D	C	D	A
42)	D	B	A	A	B
43)	B	B	C	D	C
44)	B	B	C	C	B
45)	C	C	B	D	A

Made in the USA
Columbia, SC
20 May 2019